Dietary Research and Cancer

Rajesh N. Gacche

Dietary Research and Cancer

Springer

Rajesh N. Gacche
Tumor Biology Laboratory, Department of Biotechnology
Savitribai Phule Pune University
Pune, Maharashtra, India

ISBN 978-981-16-6049-8 ISBN 978-981-16-6050-4 (eBook)
https://doi.org/10.1007/978-981-16-6050-4

This Springer imprint is published by the registered company Springer Nature Singapore Pte Ltd.
The registered company address is: 152 Beach Road, #21-01/04 Gateway East, Singapore 189721, Singapore

Dedicated to my parents, who besides illiterate had a courage to admit me in 'School' and made my life meaningful......!!!

Foreword

Great strides of appreciable efforts have been made in decreasing the progression of cancer; however, the global incidence and cancer-related mortality are still substantially increasing across the world. We still have to understand more about the molecular drivers of cancer progression. Sizable volume of research also pin-points at the pathophysiology of cancer with individual's lifestyle, physical activity, dietary habits and mindset. Owing to the essential role played by these contributing factors, many National Health Programmes and world reputed organizations have taken initiatives in providing guidance, support and evidence-based strategies for the cancer survivors, health service providers and populations for reducing the cancer risk. In fact, such endeavours for providing evidence-based strategic approaches are more authentic when there is a collaborative network of experts from the area of cancer biology, epidemiology, prevention, public health, policy makers and health service providers, especially oncologists and medical professionals.

The advances in cancer biology research have clearly demonstrated that the progression of tumour growth is closely associated with the supply of nutrients that are usually derived from the dietary inclusions of the host. These understandings further paved the way for modifying the individual's diet that may alter the nutrient availability in the tumour microenvironment and may target the metabolic vulnerabilities of the cancer cells and might also enhance the efficacy of cancer chemotherapy drugs. Thus, dietary interventions at present represent a promising strategic approach for integrated management of cancer.

Tailoring dietary approaches has remained one of the important treatment options for the management of obesity. The evolving literature links obesity with the onset of diabetes. The current molecular understandings shed light on diabetes-related factors which are associated with risk, progression, mortality and treatment of cancer. At present, diabetes seems to be a neglected risk factor for the neoplastic transformation and only limited literature is available on their correlation, and perhaps the lack of mechanistic understanding in such comorbid conditions may contribute to higher mortality rates. Owing to glucose addiction of cancer cells, the future research settings have an opportunity to understand the prevalence of hyperglycaemia–cancer. Dietary interventions are also known to modulate the diabetes-related cytokines such as adiponectin, resistin and leptin. The recent findings have unravelled the role of resistin in the development of insulin resistance under obese condition, and it is also linked with a variety of human ailments including cancer. There is equal scientific opportunity for understanding the effect of different dietary patterns on resistin regulation as recent reports suggest that the secretion of resistin in rodents is strongly influenced by genetics as well as dietary interventions.

Besides the physiological functions of cholesterol, the dietary sources that contain high amounts of cholesterol have been attributed with a variety of human degenerative diseases and disorders like obesity, cardiovascular disease, diabetes and cancer. More precisely, a series of clinical observations revealed the dynamics of blood cholesterol levels in cancer patients. The scientific endeavours towards understanding how the cholesterol is being inscrutably hijacked for the development of malignancy, progression, survival, stemness and metastasis of cancerous cells will be of immense importance for modifying the dietary patterns and also for tailoring novel therapeutic approaches for the treatment of cancer. Of note, the dysregulation of the cholesterol metabolism has been shown to hamper various treatment strategies in different types of human cancers.

The book "Dietary Research in Cancer" by Prof. Rajesh Gacche is a landscape of information related to "Diet-Cancer" discourse touching multiple facets of diet-related mechanisms in the pathophysiology of cancer progression. Besides the dietary research in a variety of cancers, Prof. Gacche has also taken due care of providing diet-related information to cancer survivors. The dietary recommendations from worldwide cancer-related organizations have been presented in this book, which perhaps will be a ready reference booklet for cancer patients. The book can be an exhaustive resource for the researchers working in the mainstream of "Diet-Nutrition-Cancer". Several recent advances like the effect of dietary interventions/ patterns on cancer stem cells, oncogenic and tumour suppressor miRNAs, cancer epigenetics, anticancer drugs targets, etc. have been precisely consolidated in this book by citing current literature. Prof. Gacche has also made an attempt to present clinical evidence-based answers to a variety of diet-related questions and illusions. More precisely, the chapters on fasting/calorie restrictions in concert with chemotherapy efficiency and cancer prevention clearly outlines the current state affairs in cancer progression and diet.

I am sure that this book will be a ready reference in providing resourceful information to the researchers, healthcare professionals, oncologists, policy makers, cancer survivors, nursing staff and even to general public for maintaining a diet-related healthy profile which perhaps will be a value addition towards the management of cancer-associated complication. I convey my best wishes to Prof. Gacche for his future endeavours.

National Centre for Cell Science, Pune, Maharashtra, India

Manoj Kumar Bhat, PhD, FNASc

Preface

The clinical report of cancer diagnosis is a shocking news and perhaps from patient's point of view is a "DARK DAY" for any affected individual and his/her family and relatives. Cancer is a chronic disease and the emotional and physical burden many times be overshadowed by the financial burden incurred by the family of cancer patients. Mostly in developed countries, a major part of the financial burden is shared by the state agency. But even so, the reports from the developed countries state that cancer patients and families have to bear a significant financial load besides several functional National policies and plans that are developed to circumvent this issue. Therefore, the word "financial toxicity" has become familiar in the mainstream of cancer chemotherapy drugs owing to the progressive high prices of newer classes of effective anticancer drugs.

Apart from the real classical disease-causative factors and physiological imbalances, the other factors like social, cultural and economic also contribute significantly towards the onset and management of a variety of human diseases including cancer. Diet and nutrition have remained an important part of cancer management. Availability of appropriate kinds of foods to cancer patients during and after the cancer treatment has been proved to be helpful in effective management of cancer. Series of diet and nutrition-related questions raised by cancer survivors are waiting to be addressed scientifically. Diet and nutritional recommendations and guidelines suggested by worldwide agencies are either inconsistent or lack the mechanistic and rigorous clinical evidence. Patient-oriented and cancer-specific dietary recommendations are limited. Moreover, the impact of dietary recommendations is not effective owing to moderate interest of clinical oncologists in diet and nutritional interventions. The evolving preclinical and clinical research suggests that the survival of cancer patients is closely associated with dietary interventions, pattern of diet and role of dietary ingredients in the pathophysiology of cancer progression. The cancer survivors have altered physiology owing to the serious side effects and off-target toxicities of chemo- and radiotherapy regimes. Moreover, the toxic effects of therapeutic regimes are more devastating and worst in malnourished or undernourished cancer survivors; therefore, the convincing dietary and nutrition guidelines before, during and after treatment needs to be formulated in concert with different types of cancer and pathophysiology of cancer progression. In the current state-of-the-art, while maintaining the healthy status of cancer patients through healthy diet, it is important to consider whether the dietary interventions have any role to play in the framework of tumour/cancer metabolism and thereby affecting the cancer landscape by modulating a variety of molecular underpinnings like cell proliferating signalling pathways, epigenetic factors, metastasis, cancer stem cells, tumour suppressor or oncogenic miRNAs, and interaction with a variety of anticancer drug targets. The advances in preclinical and clinical research have crossed the general notion of "eat healthy–be fit–reduce cancer risk" and a new paradigm of evidence-based personalized dietary interventions are evolving. Overall, the diet and nutrition management is coming up as a complementary or adjuvant therapeutic approach for the effective amelioration of cancer. Till date, FDA has not approved any dietary intervention for cancer prevention. Oncologists, clinicians, dieticians and professional cancer care taker teams are expected to encourage evidence-based dietary recommendations for cancer survivors. This book offers an updated information and discourse about various dimensions of dietary research in cancer such as comparative profile of dietary recommendations by National and International agencies, frequently asked diet and nutrition-related questions by the cancer survivors and possible answers, pros and cons of dietary patterns, outcome of diet-related clinical trials, emerging framework of molecular mechanisms and interactions of dietary interventions and their applications in personalized diet, ground realities of benefits and regulatory framework for functional foods, nutraceuticals and supplements in cancer prevention and upcoming future prospectus in diet-cancer research. An elaborative discourse on cancer and fasting (including calorie restriction and fasting mimicking diet) has been described with possible answers to questions like can calorie restriction help fight cancer? Does it improve the therapeutic index of anticancer drugs and immunosurveillance? Is there valid clinical evidence of fasting/calorie restriction/fasting mimicking diet-driven improvements in overall survival, chemotherapy efficacy and drug toxicity? Looking forward, the most upcoming discourse may be to investigate the dynamic interplay between dietary interventions and cancer microbiome. There is ample research opportunity to understand the cancer-specific role of human microbiome. Moreover, the future research opportunities in understanding the role of dietary intervention in improving cancer immunosurveillance may be useful in integrating these nonpharmacological regimens with cancer immune therapy.

I hope the information described in this book may serve as a reference guideline for oncologist, physicians, health cares, food industry, cancer care takers, nursing and other stockholders while designing the treatment strategies for the effective management of cancer.

Pune, Maharashtra, India Rajesh N. Gacche

Acknowledgement

At the outset, let me express my sincere thanks to Dr. Manojkumar Bhat (Ph.D. FNASc, India), Director and a Senior Scientist, National Centre for Cell Science (NCCS), Pune, for forwarding this book to the researcher community. I would like to express my sincere thanks and gratitude to my Ph.D. and post-doctoral students of my laboratory, especially Dr. Navnath Kumbhar, Dr. Sourabh Zimare, Mr. Shridhar Chougule, Dr. Sonali Kamble, Dr. Upasana Kapoor, Miss. Pallavi Shewale, Miss, Varsharani Merugu, Miss. Snehal Nimal, Mr. Rushikesh Pol and Mr. Sharad Tat in organizing various aspects of this book. The figures and graphical illustrations in this book are original and 'Created with BioRender.com' using a purchased software copy.

Contents

About the Author

Rajesh N. Gacche, PhD, LLB is currently working as a Prof. and Head of the Department of Biotechnology of Savitribai Phule Pune University (A National Center of DBT, New Delhi, Govt. of India). He has 26 years of teaching and research experience in Biotechnology. Owing to his excellent track record of research publications, Prof. Rajesh has got a place in "Top 2% Scientist of the World—2020" (a database by Stanford University, USA). His laboratory data on "Seroprevalence and Risk Factors for Hepatitis-C Virus Infection" has got a place in "Global Hepatitis Report 2017" published by the World Health Organization in 2018. Prof. Rajesh has published over 115 research papers in the area of "Biomedical Research" in reputed International journals with impact factor range of 1–18.50, h-index of 29 and citations over 2600. He has published his work in prestigious *Nature-Oncogenesis* (Nature Publications, IF 7.4) and many other highly reputed journals like *Drug Resistance Updates* (Impact Factor 18.50) and *BBA Cancer* (Impact Factor 10.68). His lab has completed one dozen Extramural Major Research Projects funded by various funding agencies of India with four patents to his credit and 15 students awarded Ph.D. under his guidance.

We are in the midst of "target rich lead poor scenario" as the evolving drug targets are not being compensated by the novel and effective leads. In the current state-of-the-art, over 60% anticancer drugs that are prescribed by the physician are basically derived from the natural products mostly from plants, marine organisms and microorganisms. The anticancer activity of most natural products often acts via regulating cell proliferation pathways, immune function, inducing apoptosis or autophagy, or inhibiting cell proliferation. Prof. Rajesh's lab works on identification of evidence-based novel leads from natural sources (including dietary sources) and being validated against several anticancer drug targets with possible molecular mechanisms. In parallel to this, his lab also focuses on addressing the basic molecular aspects of Cancer Biology in relation to drug resistance, epigenetics, heterogeneity, angiogenesis, EMT and apoptosis. His lab also works on applications of computational tools for defining SAR, designing peptide vaccine and drug repurposing. His research on the role of dietary flavonoids (and medicinal herbs) in the management of human degenerative diseases, especially cancer and diabetic cataract, has been published in several International reputed journals.

Prof. Rajesh has been bestowed with several awards like Best Teacher Award, "Fellow of Maharashtra Academy of Science" Young Scientist Award of SERC DST, New Delhi. He is a member of National Academy of Science (NASI), India. He has received "Award of Appreciation" by Italian Ministry of Health, Rome, Italy, for his contribution to the evaluation of the research projects presented with reference to Biomedical Call for Grant from 2015 to 2019; certificate of "Award of Appreciation—2016" by National Science Foundation (NSF), Govt. of Croatia, for his contribution to the evaluation of the research projects submitted to NSF, Govt. of Croatia; certificate of "Highly Cited Research" in Progress in Biophysics and Molecular Biology, awarded in December 2016 by Elsevier Publisher and "An award of Appreciation for Research in Cancer Biology" presented in Global Biotechnology Congress—2014 at Hynes Convention Center Boston, Boston, USA, in June 2014. Prof. Rajesh visited countries like the USA, Austria, China, Singapore, etc. and chaired International Conference Sessions and delivered lectures in several National and International Conferences/Symposia.

Chapter 1
Introduction to Dietary Research and Cancer

Abstract Epidemiological studies have clearly demonstrated the fact that diet and nutrition have profound impact and influence on the progression of cancer and also on risk of developing cancers. Besides great strides of efforts in dietary research, till today series of diet- and nutrition-related questions raised by cancer survivors are waiting to be addressed scientifically. Diet and nutritional recommendations and guidelines suggested by worldwide agencies are either inconsistent or lack the mechanistic and rigorous clinical evidence. Patient-oriented and cancer-specific dietary recommendations are limited. Moreover, the impact of dietary recommendations is not effective owing to moderate interest of clinical oncologists in diet and nutritional interventions. The evolving preclinical and clinical research suggests that the survival of cancer patients is closely associated with dietary interventions, pattern of diet, fasting, calorie restrictions and action of dietary ingredients with respect to the pathophysiology for cancer progression. Furthermore, the outcome of preclinical research has demonstrated that the prognosis of cancer could be altered by modifying or restricting diet. The emerging research in diet–cancer discourse includes the diet/metabolite-mediated regulation of cancer signalling pathways/ growth factors, interaction with anticancer targets, metastasis, cancer stem cells, cancer-related miRNAs, drug–diet interaction, immunosurveillance, drug toxicity, side effects and cancer microbiome. The upcoming trends in the dietary research are coevolving with the advances in molecular aspects of cancer biology, and perhaps such research outcome may serve as a baseline reference for developing evidence-based guideline for various stakeholders in the mainstream of cancer management.

Keywords Dietary guidelines · Cancer patients · Dietary interventions · Cancer progression · Dietary supplements · Side effects · Toxicity · Fasting · Chemotherapy

Introduction

The GLOBOCAN 2020 report on cancer and allied mortality across the world published by the International Agency for Research on Cancer has distinctly stated the figures of serious concern of various types of cancer worldwide. As per the

R. N. Gacche, *Dietary Research and Cancer*,
https://doi.org/10.1007/978-981-16-6050-4_1

report, "worldwide, a predicted 19.3 million people developed novel cancer cases (18.1 million except nonmelanoma skin cancer) and an estimated 10.0 million deaths (9.9 million except nonmelanoma skin cancer) occurred due to cancer in the year 2020. Women's breast cancer is found to be the most commonly diagnosed cancer with 2.3 million novel cases (11.7%) accompanied by lung (11.4%), colorectal (10.0%), prostate (7.3%) plus stomach cancer (5.6%). Lung cancer has always been a leading cause of cancer linked deaths with a predicted 1.8 million deaths (18%) accompanied by colorectal (9.4%), liver (8.3%), stomach (7.7%) along with female breast cancer (6.9%). The general incidence ranges from two to three times higher in transitioned vs. transitioning nations in case of both sexes, while mortality was diverse like <2-fold for men and less for women. The mortality rate of breast and cervical cancers, nevertheless, was significantly greater in transitioned vs. transitioning nations (15.0 vs. 12.8 per 100,000 and 12.4 vs. 5.2 per 100,000 accordingly). The global cancer trouble is anticipated to be 28.4 million cases by 2040, an increase of 47% from 2020, with a significant increase in transitioning (64–95%) compared to transitioned countries (32–56%) as a result of demographic variations, even though this can also be aggravated by the rising risks allied with globalization and the increasing economy" (Sung et al. 2021). This report, while outlining the incidence and mortality of various cancers around the world, specifically highlighted the impact of diet, diet pattern and nutrition on the risk factors associated with various cancer deaths and deaths worldwide (Sung et al. 2021). A current study from the USA has demonstrated that over 80,000 new cancer cases (5.2% of all novel cancer cases in 2015) were allied with the ingestion of suboptimal diet and colorectal cancer had the maximum number and percentage of diet-associated cases (Zhang et al. 2019a, b).

The cancer survivors have altered physiology owing to the serious side effects and off-target toxicities of chemo- and radiotherapy regimes. Moreover, the toxic effects of therapeutic regimes are more devastating and worst in malnourished or undernourished cancer survivors; therefore, the convincing dietary and nutrition guidelines before, during and after treatment need to be formulated in concert with different types of cancer and pathophysiology of cancer progression. In the current state of the art, while maintaining the healthy status of cancer patients through healthy diet, it is important to consider whether the dietary interventions have any role to play in the framework of tumour/cancer metabolism and thereby affecting the cancer landscape by modulating variety of molecular underpinnings like cell proliferating signalling pathways, epigenetic factors, metastasis, cancer stem cells, tumour suppressor or oncogenic miRNAs, and interaction with variety of anticancer drug targets. The advances in preclinical and clinical research have crossed the general notion of "eat healthy-be fit-reduce cancer risk", and a new paradigm of evidence-based personalized dietary interventions are evolving. Overall, the diet and nutrition management are coming up as a complementary or adjuvant therapeutic approach for the effective amelioration of cancer. Till date, FDA has not approved any dietary intervention for cancer prevention. Oncologists, clinicians, dieticians and professional cancer caretaker teams are expected to encourage evidence-based dietary recommendations for cancer survivors. The landscape of dietary research includes multiple issues such as development of evidence-based convincing dietary recommendations by national and international agencies, providing scientific answers to frequently asked diet- and nutrition-related questions by the cancer survivors, pros and cons of dietary patterns, outcome of diet-related clinical trials, emerging framework of molecular mechanisms and interactions of dietary interventions and their applications in personalized diet, ground realities of benefits and regulatory framework for functional foods, nutraceuticals and supplements in cancer prevention and upcoming future prospectus in diet–cancer research. Series of clinical trials have been completed, and few more are in progress investigating the benefits of fasting/calorie restriction/fasting-mimicking diet in cancer prevention. The research is also expanding towards assessing the effect of fasting/calorie restriction/fasting mimicking in improving the therapeutic index of anticancer drugs and response to chemotherapy. The upcoming trends in the dietary research are coevolving with the advances in molecular aspects of cancer biology, and perhaps such research outcome may serve as a reference for developing evidence-based guideline for oncologists, physicians, food industry, cancer caretakers, nursing and other stockholders while designing the treatment strategies for the effective management of cancer.

The Evolving Trends in Dietary Research

"Diet and nutrition" are the buzzwords in the mainstream of growing health awareness across the world, and it is equally applicable for the management of degenerative diseases including cancer. The well-known quote that "let food be thy medicine and medicine be thy food" by Hippocrates still holds a contemporary significance even after 25 centuries. In a very traditional notion, human diet is considered as a double-edged sword that may influence both causation and prevention of cancer, per say these traditional claims seem to be valid to a certain extent, as the epidemiological and laboratory evidence describes certain dietary components as potential agents with increased risk of developing cancer (Vineis and Wild 2014; WCRF: Expert Report 2018), while others are implicated as cancer-preventive agents (Maruca et al. 2019). In fact, the majority of the prominent hypotheses that have been framed for understanding the effects of diet on cancer risk and patient survival have been derived from observations related to dietary patterns and cancer incidence rates in different populations across the world. In human body, one of the most frequently encountered factors by cancer cells is the diet and nutritional ingredients, which may influence the cancer cell microenvironment and tumour metabolism by acting as positive or negative regulators of cancer progression (Maddocks et al. 2017; Sullivan et al. 2019). There is evolving research interest in using dietary modification to complement standard cancer therapy; however, the currently formulated dietary recommendations are not standardized and clinically validated. Nevertheless, many of such putative recommendations/guidelines have not been well integrated with clinical practices. There is ample scope and opportunity in constituting an optimal diet for cancer patient (The Cancer Research UK-Ludwig Cancer Research

Nutrition and Cancer Prevention Collaborative Group 2019), by asking series of rational questions in the framework whether dietary composition/ingredients are affecting the tumour growth and cancer progression? Is there any impact of dietary alterations with availability of metabolites to tumours/cancer cells? Do diet-induced modifications in nutrient access have any effect on the landscape of tumour metabolism? Do dietary modifications are associated with metabolic perturbations or drug synergies? (Lien and Heiden 2019).

There are several hundreds of organizations functional across the world, but the agencies like WCRF-AICR (World Cancer Research Fund-American Institute for Cancer Research), ACS (American Cancer Society), NCI (National Cancer Institute), CSC (Cancer Support Community), CCS (Canadian Cancer Society), WHO (World Health Organization), ECAC (European Code Against Cancer), NCCN (National Comprehensive Cancer Network), ESPEN (The European Society for Clinical Nutrition and Metabolism) CRUK (Cancer Research UK) are few of the dedicated organizations providing diet- and nutrition-related information and dietary guidelines/recommendations to cancer patients. Among the aforesaid agencies, the WCRF-AICR and ACS have developed cancer-specific dietary guidelines based on evidence-based dietary interventions. WCRF-AICR and ACS recommendations have been also validated in several large prospective epidemiological settings, which have demonstrated significant decline in the prevalence of cancer in a range of 17% to over 50% and significant decrease in cancer-related mortality by 20–30%. Several reports describe that the systematic implementation of WCRF-AICR and ACS guidelines has significant impact in reducing the cancer risk, mortality and improvement in survival and overall health of cancer patients (Vergnaud et al. 2013; Kabat et al. 2015; Jankovic et al. 2017). Although there are no consistency and consent on many dietary issues, however, the information, guidelines and series of updated contemporary research studies are ready references for cancer survivors in planning the diet and nutrition before, during and after the cancer treatment. In fact, the dietary and nutrition guidelines need to be integrated with clinical practice; however, due to the lack of dietary interest in oncologists, physicians and patient awareness, the global implementation of cancer diet guidelines is a challenging issue.

Overall, plethora of preclinical, clinical, epidemiological, cohort and observational studies of couple of decades have established sound scientific grounds that highlight the significance of dietary and nutritional interventions as key players in causation and progression of variety of human degenerative diseases including cancer (Global Nutrition Report, 2019). Sizable literature has accumulated in the recent past, which has crossed the conventional preclinical settings of addressing anticancer effects of dietary metabolites. The current literature describes the mechanistic and target-oriented effect of dietary metabolites unravelling the up- and downregulation of variety of physiological pathways implicated in cancer progression, anticancer drug targets, cancer stem cells, oncogenic or tumour-suppressing microRNAs (miRNAs) (Tan et al. 2011; Singh et al. 2017; Maruca et al. 2019; Samec et al. 2019). With these molecular progressions in the mainstream of "diet/nutrition–disease" research in association with pathophysiology of cancer progression, there has been a paradigm shift in understanding the effect of diet on cancer progression, survival and mortality. Perhaps the evolving research in this direction might inspire the food and nutrition industry for design and development of effective personalized, cancer-specific dietary choices.

Apart from general anticancer activity of dietary metabolites (Maruca et al. 2019), the mechanistic target-oriented molecular underpinnings have opened several new avenues of combating degenerative diseases using dietary interventions, which perhaps will serve as a guideline for oncologist. For example, an evolving research window of dietary metabolite-driven epigenetic alterations in the landscape of cancer has attracted the attention of research community as these effects are directly associated with the genetic basis of disease progression (Hardy and Tollefsbol 2011). It is well-established fact that there is an aberrant gene expression profile owing to epigenetic alterations in all cancer types; therefore, consumption of bioactive dietary ingredients will be an intuitive and holistic approach for cancer management, which will help to stabilize the epigenomic cancer aberrations and may also serve as a complementary therapeutic approach for the effective cancer treatment (Dawson and Kouzarides 2012; Lewis and Tollefsbol 2017). Casting the epigenetic diet for cancer patient will be an exciting endeavour as currently there are seven epigenetic drugs that inhibit epigenetic writers and erasers approved by US FDA for the treatment of cancer (Duan et al. 2019). Perhaps the future dietary settings and food and nutrition industry might use an annotation of "Epigenetic Diet for Cancer Patient" in general and personalized cancer-specific epigenetic diet in particular.

The past decade research has also fuelled one more research window with a significant hope of novel therapeutic approach that links the dietary interventions with regulation of series of oncogenic and tumour suppressor miRNAs (Samec et al. 2019). miRNAs have been identified as small noncoding RNAs that act as post-transcriptional regulators of gene expression and thereby influence the pathophysiology of disease progression. Series of dietary bioactive compounds like curcumin, apigenin, epigallocatechin-3-gallate (EGCG), resveratrol, quercetin, palmatine, folate, genistein, ellagic acid and indole-3 carbinol are implicated in regulation of oncogenic and tumour suppressor miRNAs (Gezici and Sekeroglu 2017; Banikazemi et al. 2018).

To a greater extent, the cancer mortality is closely associated with metastasis. On the eve of evolving cancer heterogeneity and drug resistance, physicians and oncologist need to consider the dietary interventions as they are influential in modulating the process of metastasis. The majority of the cancer chemotherapy drugs are targeted towards the primary cancer cells, and in the current situation, there are no tailored drugs that target metastasis as a whole. Revisiting the diet and dietary patterns of cancer patients is advisable, as the evolving preclinical research and clinical observations focus on the importance of diet in concert with cancer metastasis. For example, a recent preclinical study has demonstrated the positive effect of high-fat diet on prostate cancer metastasis (Chen et al. 2018); on the other hand, the polyphenols present in green tea have been reported to hinder variety of metastatic targets in liver, lung, prostate, skin, breast and gastrointestinal cancer in preclinical studies (Khan and Mukhtar 2010). To a greater extent, the physiology of cancer patients who are surviving in metastasis stage is worst. In general, the metastatic cancers, also called as progressive or advanced cancer, are unlikely to be cured; under such circumstances, the focus of treatment modalities shifts from conventional treatments to emotional, physical and practical ways of extending patients survival using alternative therapies. The research findings suggest that selecting evidence-based dietary factors might help in extending the survival of cancer patient. Series of dietary factors and micronutrients have been reported to act on metastatic targets and metastasis-suppressing genes in preclinical model studies (Meadows 2012; Niedzwiecki et al. 2010).

Diet has remarkable influence for tissue homeostasis, human physiology and health. Diet-driven proliferation or inhibition of cancer stem cells is not an exception to this physiological notion (Li et al. 2019). A better understanding of diet-driven metabolic pathways and their impact on tissue homeostasis, cancer commencement and cancer stem cell regulation might pave a way for designing therapeutic dietary inclusions for cancer patients, as nonjudicious inclusions of dietary interventions may boost stem cell numbers, which perhaps may impose risk of cancer recurrence. A state-of-the-art information on how dietary interventions control metabolism in concert with cancer stem cell functions has been reviewed elsewhere (Mihaylova et al. 2014). Moreover, the recent preclinical model studies clearly demonstrated the effect of dietary metabolites on regulation of cancer stem cells. For example, high-fat diet (HFD) has been shown to have dramatic effect on biology of stem cells and it has been reported that HFD improves the stemness and tumorigenicity of intestinal cancer cells (Beyaz et al. 2016). Intake of HFD leads to obesity, and the preclinical study clearly demonstrated that the obesity develops colonic stem cell expansion through cancer initiation (DeClercq et al. 2015). Although the cancer stem cell research in relation to diet is in infancy, however, the upcoming scientific cues are helpful for designing dietary interventions for cancer patients and a reference for physicians, oncologists and caretaker teams.

Looking towards the future prospectus of diet–cancer discourse, the landscape of dietary research may be extended to several dimensions of anticancer research. For example, the most upcoming discourse may be to investigate the dynamic interplay between dietary interventions and cancer microbiome. There is ample research opportunity to understand the cancer-specific role of human microbiome. Moreover, the future research opportunities in understanding the role of dietary intervention in improving cancer immunosurveillance may be useful in integrating these nonpharmacological regimens with cancer immune therapy.

Conclusion

The upcoming trends in the dietary research are coevolving with the advances in molecular aspects of cancer biology and perhaps such research outcome may serve as a reference for developing evidence-based guideline for oncologists, physicians, food industry, cancer caretakers, nursing and other stockholders while designing the treatment strategies for the effective management of cancer. The emerging research in diet–cancer discourse includes the diet/metabolite-mediated regulation of cancer signalling pathways/growth factors, interaction with anticancer targets, metastasis, cancer stem cells, cancer-related miRNAs, drug–diet interaction, immunosurveillance, drug toxicity, side effects and cancer microbiome. The upcoming trends in the dietary research are coevolving with the advances in molecular aspects of cancer biology, and perhaps such research outcome may serve as a baseline reference for developing evidence-based guideline for various stakeholders in the mainstream of cancer management. More and more diet-related clinical trials investigating the role of dietary intervention (dietary factors, bioactive molecules, patterns, fasting, etc.) in cancer prevention, improving chemotherapy, overall survival and quality of life and reducing toxicity of chemodrugs are required for effective implementation of dietary interventions in clinical practice.

References

Banikazemi Z, Haji HA, Mohammadi M, Taheripak G, Iranifar E, et al. Diet and cancer prevention: dietary compounds, dietary MicroRNAs, and dietary exosomes. J Cell Biochem. 2018;119 (1):185–96.

Beyaz S, Mana MD, Roper J, Kedrin D, Saadatpour A, Hong S-J, et al. High-fat diet enhances stemness and tumorigenicity of intestinal progenitors. Nature. 2016;531(7592):53–8.

Chen M, Zhang J, Sampieri K, Clohessy JG, Mendez L, Gonzalez-Billalabeitia E, Liu XS, et al. An aberrant SREBP-dependent lipogenic program promotes metastatic prostate cancer. Nat Genet. 2018;50(2):206–18.

Dawson MA, Kouzarides T. Cancer epigenetics: from mechanism to therapy. Cell. 2012;150:12–27.

DeClercq V, McMurray DN, Chapkin RS. Obesity promotes colonic stem cell expansion during cancer initiation. Cancer Lett. 2015;369:336–43.

Duan YT, Sangani CB, Liu W, Soni KV, Yao Y. New promises to cure cancer and other genetic diseases/disorders: epi-drugs through epigenetics. Curr Top Med Chem. 2019;19(12):972–94. https://doi.org/10.2174/1568026619666190603094439.

Gezici S, Sekeroglu N. Regulation of MicroRNAs by Natural products and bioactive compounds obtained from common medicinal plants: novel strategy in cancer therapy. Indian J Pharm Educ Res. 2017;51(3 Suppl):S483–8.

Hardy TM, Tollefsbol TO. Epigenetic diet: impact on the epigenome and cancer. Epigenomics. 2011;3(4):503–18.

Jankovic N, Geelen A, Winkels RM, et al. Adherence to the WCRF/AICR dietary recommendations for cancer prevention and risk of cancer in elderly from Europe and the United States: a meta-analysis within the CHANCES Project. Cancer Epidemiol Biomark Prev. 2017;26:136–44.

Kabat GC, Matthews CE, Kamensky V, et al. Adherence to cancer prevention guidelines and cancer incidence, cancer mortality, and total mortality: a prospective cohort study. Am J Clin Nutr. 2015;101:558–69.

Khan N, Mukhtar H. Cancer and metastasis: prevention and treatment by green tea. Cancer Metastasis Rev. 2010;29:435–45.

Lewis KA, Tollefsbol TO. The influence of an epigenetics diet on the cancer epigenome. Epigenomics. 2017;9(9):1–3.

Li W, Zimmerman SE, Peregrina K, Houston M, Mayoral J, Zhang J, et al. The nutritional environment determines which and how intestinal stem cells contribute to homeostasis and tumorigenesis. Carcinogenesis. 2019;40(8):937–46.

Lien EC, Vander Heiden MG. A framework for examining how diet impacts tumour metabolism. Nat Rev Cancer. 2019 Nov;19(11):651–61.

Maddocks ODK, Athineos D, Cheung EC, Lee P, Zhang T, van den Broek NJF, et al. Modulating the therapeutic response of tumours to dietary serine and glycine starvation. Nature. 2017;544 (7650):372–6.

Maruca A, Catalano R, Bagetta D, Mesiti F, Ambrosio FA, Romeo I, Moraca F, Rocca R, Ortuso F, Artese A, Costa G, Alcaro S, Lupia A. The Mediterranean Diet as source of bioactive compounds with multi-targeting anti-cancer profile. Eur J Med Chem. 2019;31(181):111579.

Meadows GG. Diet, nutrients, phytochemicals, and cancer metastasis suppressor genes. Cancer Metastasis Rev. 2012;31(3–4):441–54.

Mihaylova MM, Sabatini DM, Yilmaz OH. Dietary and metabolic control of stem cell function in physiology and cancer. Cell Stem Cell. 2014;14:292–305.

Niedzwiecki A, Roomi MW, Kalinovsky T, Rath M. Micronutrient synergy—a new tool in effective control of metastasis and other key mechanisms of cancer. Cancer Metastasis Rev. 2010;29:529–42.

Samec M, Liskova A, Kubatka P, Uramova S, Zubor P, Samuel SM, et al. The role of dietary phytochemicals in the carcinogenesis via the modulation of miRNA expression. J Cancer Res Clin Oncol. 2019;145(7):1665–79.

Singh AK, Sharma N, Ghosh M, Ho YP, Jeong DK. Emerging importance of dietary phytochemicals in fight against cancer: role intargeting cancer stem cells. Crit Rev Food Sci Nutr. 2017;57:16.

Sullivan MR, et al. Quantification of microenvironmental metabolites in murine cancers reveals determinants of tumor nutrient availability. Elife. 2019;8:e44235.

Sung H, Ferlay J, Siegel RL, Laversanne M, Soerjomataram I, Jemal A, Bray F. Global cancer statistics 2020: GLOBOCAN estimates of incidence and mortality worldwide for 36 cancers in 185 countries. CA Cancer J Clin. 2021;71(3):209–49. https://doi.org/10.3322/caac.21660.

Tan AC, Konczak I, Sze DMY, Ramzan I. Molecular pathways for cancer chemoprevention by dietary phytochemicals. Nutr Cancer. 2011;63(4):495–505.

The Cancer Research UK-Ludwig Cancer Research Nutrition and Cancer Prevention Collaborative Group. BMC Med. 2019;17(148):1–15.

Vergnaud AC, Romaguera D, Peeters PH, et al. Adherence to the World Cancer Research Fund/ American Institute for Cancer Research guidelines and risk of death in Europe: results from the European Prospective Investigation into Nutritionand Cancer cohort study1,4. Am J Clin Nutr. 2013;97:1107–20.

Vineis P, Wild CP. Global cancer patterns: causes and prevention. Lancet. 2014;383:549–57.

WCRF (World Cancer Research Fund)/American Institute for Cancer Research. Continues Update Project Expert Report 2018. Recommendations and public health and policy implications. 2018. dietandcancerreport.org. Accessed 24 Sept 2020.

Zhang FF, Cudhea F, Shan Z, et al. Preventable cancer burden associated with poor diet in the United States. JNCI Cancer Spectr. 2019a;3(2):pkz034.

Zhang X, Tang T, Pang L, Sharma SV, Li R, Nyitray AG, Edwards BJ. Malnutrition and overall survival in older adults with cancer: a systematic review and meta-analysis. J Geriatr Oncol. 2019b;10(6):874–83.

Chapter 2
Healthy Dietary Patterns and Cancer

Abstract The research in the past two decades identified the dietary patterns and bioactive ingredients, which helped to establish a baseline data for the formulation of cancer prevention guidelines and fuelled the area of developing dietary patterns with the aim of cancer prevention and improving the overall survival of cancer patients. Conventionally, a dietary pattern is described as "The quantity, variety, or combination of different foods and beverage in a diet and the frequency with which they are habitually consumed". Among the dietary pattern's variety, Mediterranean pattern (which comprises of olive oil, fruits, legumes, fish and vegetables), ketogenic dietary pattern (a high-fat intake, moderate protein and low-carbohydrate intake), vegan (abstains from all animal products) and vegetarian dietary pattern, alkaline dietary pattern and specific diet (deficient of one or more unique molecules) are some of the dietary patterns associated with cancer prevention. The preclinical and clinical data accumulated in the recent past advocate the benefits of different dietary patterns in relation to different cancers; however, there is little convincing evidence, which suggests that particular dietary pattern positively influences the outcomes in relation to prognosis, recurrence and mortality. The limited evidences exist, which describe the individual dietary patterns and their useful effects; however, overall, the Western dietary pattern seems to be detrimental for variety of cancers.

Keywords Dietary patterns · Cancer · Ketogenic diet · Mediterranean diet · Alkaline diet · Western diet · Clinical risk and benefits

Abbreviations

ACS	American Cancer Society
AICR	American Institute for Cancer Research
KD	Ketogenic diet
MediD	Mediterranean diet
MR	Methionine-restricted
WCRF	World Cancer Research Fund

R. N. Gacche, *Dietary Research and Cancer*,
https://doi.org/10.1007/978-981-16-6050-4_2

Introduction

Malnutrition in cancer patients is one of the serious issues that need to be linked with dietary interventions. With progressive advancement of cancer, the malnutrition rate increases as a result of energy demands of cancer cells are increased; moreover, nutritional intake is reduced. Malnutrition is a common observation in majority of the hospitalized older adult cancer patients and is strongly associated with variety of complications like fat and muscle wasting, poor response to treatment modalities and thereby increased morbidity and mortality (Zhang et al. 2019a, b; Na et al. 2018). A recently published 1588 hospitalized patients' prospective cohort study revealed that malnutrition-related complications are more severe in cancer patients having lung or pancreaticobiliary and oesophageal cancer when compared with colon, liver or stomach cancer. Of note, patients who undergo chemo-, radiotherapy or supportive concern are more susceptible to malnutrition as compared to patients undergoing surgery (Na et al. 2018). Perhaps, proper dietary management may help in increasing the tolerance to treatment regimens and related outcomes. As such, there is no superfood kind of dietary interventions, which will significantly contribute towards cancer prevention and patient survival; moreover, aspiring for such a single superfood will be a mere illusion. Perhaps, a holistic dietary approach is likely to be most beneficial for keeping the good health of cancer survivors.

Owing to the accumulation of preclinical and clinical evidence on the effect of dietary patterns on amelioration of chronic human ailments, the significance of dietary patterns has now got place in the US DGA (US Department of Health and Human Services, US Department of Agriculture 2015). This is particularly worthy of noting because the interactive and additive effects of different dietary patterns might contribute towards the cancer prevention (World Cancer Research Fund/American Institute for Cancer Research. Continues Update Project Expert Report 2018). The evolving literature consistently supports the recommendations about following healthy dietary patterns for the effective management of cancer prevention and as adjuvant anticancer treatment modality. The Dietary Guidelines Scientific Report (2015) have summarized that there is no convincing evidence to support the role of dietary patterns (especially enriched in plant foods, low in animal-derived products and processed carbohydrates) in lowering the risk of breast cancer (postmenopausal). The report also mentioned that plant origin patterns, which are low in red (processed meat) and extra (added) sugars, are closely associated with a reduced risk of colorectal cancer (US Department of Agriculture, Dietary Guidelines Advisory Committee. Scientific Report of the 2015 Dietary Guidelines Advisory Committee 2015). Moreover, the WCRF/AICR agencies summarized that a Mediterranean dietary pattern is "convincingly" evidenced with a reduced risk of overweight, weight gain and obesity; on the other hand "Western"-type diet style is most probably concerned with a higher risk of above outcomes (World Cancer Research Fund/American Institute for Cancer Research. Continues Update Project Expert Report 2018). The Diet Patterns Methods Project, which is a multicentre collaborative study investigating the effect of dietary patterns in relation to cause-specific mortality, reported 8–17% reduced risk of cancer-related mortality among women and around 17–24% low risk in men whose diets were in agreement with scores of four healthy dietary pattern (Liese et al. 2015).

Although sizable preclinical and clinical studies link the diet/nutrition and cancer prevention with possible role in pathophysiology of cancer (Patel et al. 2018; Chikara et al. 2018), however, the convincing and conclusive clinical evidence that yields strong recommendations is still evolving and challenging issue in the mainstream of cancer and diet (Mayne et al. 2016). Moreover, establishing the link among cancer risks and dietary factors in epidemiological studies through clinical trials is circumvented with several factors. Nevertheless, the majority of randomized clinical trials and many observational studies have not considered the factors like cancer subtypes, intensity of treatment modality and associated illness. As a representative example, in a 12-year follow-up study designed for understanding the associations of quality of plant-dependent diets and mortalities regarding cancer and cardiovascular diseases, it was revealed that there are no consistent plant-dependent diet quality and cancer mortality associations and the outcome does not mount to any concrete recommendations (Baden et al. 2019). Therefore, the volume and quality of research are not full proof and sufficient to formulate the concentre dietary recommendations for cancer survivors (WCRF/AICR CUP Expert Report 2018).

There is frequent enquiry from the cancer survivors in relation to choosing dietary pattern to be adopted before, during and after cancer treatment. Besides variety of dietary patterns available across the world, in general, over 48% of cancer patients with a hope of improvement in survival and prevent recurrence have preference for some popular diets like ketogenic, alkaline, Palaeolithic, vegan, macrobiotic and Mediterranean. Of note, these dietary patterns have many positive features that are consistent with the dietary recommendations suggested by the AICR and ACS. The common factor is that all these dietary patterns advocate eating vegetables. In a KD (ketogenic diet), eating fruit is encouraged, and vegan diet discourages refined alcohol and grains, whereas most of the AICR and ACS guidelines are met by the alkaline and the macrobiotic diets however such dietary styles do not have concrete scientific evidence of their anticancer potentials, and moreover, these diets do not have inclusions of diet ingredients or food types, which have proven beneficial effects towards cancer prevention and boosting general health (Zick et al. 2018). Looking towards the current literature, it seems that there cannot be a strong dietary pattern recommendation unless otherwise the individual diet pattern is clinically proved for its efficacy against specific cancer under defined set of conditions.

Clinical Benefits of Different Dietary Patterns in Cancer Prevention

The good news is that to a greater extent the clinical evidence of few dietary patterns has started accumulating in recent years. Perhaps these clinical findings with specific cancer type may serve as a reference for oncologist and other stakeholders and an inspiration for more such studies. For example, the ketogenic diets are usually adequate-protein, low-carbohydrate and high-fat diets with a fat-to-protein and nonfibre carbohydrates' 3:4:1 macronutrient ratio. The basic aim of ketogenic diets is to compel the body for lipid metabolism rather than proteins or carbohydrates, thereby shifting the primary energy source of body to ketones from glucose. One more aspect in formulating ketogenic diets for cancer patients is the well-known Warburg effect, a phenomenon in which cancer cells are dependent on glycolysis, generating fuel for their growth even in the oxygen's presence. The advocates of the KD hypothesize that growth and proliferation of cancer cell will be adversely affected through shifting them to ketone metabolism (Weber et al. 2019). The clinical trials' outcome with KD interventions revealed the therapeutic potentials and efficacy against advanced pancreatic and lung cancer (Zahra et al. 2017), glioblastoma multiforme (van der Louw et al. 2019), primary aggressive brain cancer (Schwartz et al. 2018), ovarian and endometrial cancer (Cohen et al. 2020), breast cancer (Khodabakhshi et al. 2020), etc., and there is evolving clinical interest of integrating KD as an adjuvant cancer therapy (Klement 2019).

Series of clinical trials (completed and being recruited) for investigating the KD effect on improving chemotherapy, overall survival and quality of life of cancer patients. For example, a controlled clinical trial (NCT02516501) was designed for studying the KD effect on body composition comprising of natural food vs an unspecified standard diet in breast cancer patient that was going through radiotherapy. The outcome of the trial described that breast cancer patient going through curative radiotherapy, a KD depending on natural foods, was sufficient. The study observed that the KD tries reducing fat mass of body and weight, whereas that preserves skeletal muscle and fat-free mass after initial water losses (Klement et al. 2020). Randomized controlled trial ("Chinese Clinical Trial Registry", ID: ChiCTR1900024597) has been designed to address if irinotecan treatment in combination with KD for metastatic or recurring breast cancer in the northeast China's women patient can improve the sensitivity of disease to chemotherapy and reduce target lesions (Wang et al. 2020). A systematic review of 46 publications comprising 39 clinical studies having 770 patients revealed that there were no conclusive evidences found towards the antitumour effects or improvements in the cancer patients' overall survival. The analysis also reported that most patients with mild-to-moderate side effects undergone substantial weight loss. The study concluded that, owing to extremely heterogeneous outcomes and methodological discrepancies of the integrated studies, clinical evidence for the impact of KDs in cancer patients is still poorly understood (Römer et al. 2021). A systematic review analysis on "Diet and Chemotherapy: The Effects of Fasting and Ketogenic Diet on Cancer

Treatment" revealed that fasting and KD resulted in a reduction in side effects of adjuvant chemotherapy (perhaps because of drugs toxicity) and improved quality of life as compared to patients that follow no diet (Plotti et al. 2020).

The second dietary pattern that is evidenced with clinical benefits to cancer patients is the Mediterranean diet (MediD). MediD comprises of sufficiently balanced cocktail of polyunsaturated fats, cereals, legumes, olive oil, fish, nuts, fruit and vegetables with moderate consumption of alcohol (mostly red wine), dairy products, eggs and meat. As described by UNESCO in 2010, the significance of this famous dietary pattern lies in its health-promoting abilities and longevity. Series of clinical trials and many observational cohort studies link the MediD as protective aspect against the outbreak of several cancer types and overall health benefits (Serra-Majem et al. 2019; Mentella et al. 2019). Preclinical model studies have established the fact that dietary ingredients found in Mediterranean diet are associated with the anticancer and anti-inflammatory activities. Several clinical trial and observational cohort studies describe the role of MediD in amelioration of colorectal (Schulpen and van den Brandt 2020), prostate (Russo et al. 2019), breast (Shaikh et al. 2019), lung (Fortes et al. 2003; Schulpen and van den Brandt 2018), head and neck (Giraldi et al. 2017; Salvatore Benito et al. 2019), nasopharyngeal cancer (Turati et al. 2017), etc. In a case–control study, involving 60 hospital-based controls and 60 new diagnosed prostate cancer patients, it was observed that Western dietary pattern was related to increase in the prostate cancer risks as compared to MediD (Jalilpiran et al. 2018). Although globally MediD is advanced as the healthiest dietary patterns, its clinical impact has been limited by cancers' location, demography, culture studies, lack of positive correlation with few cancers and controversial clinical outcomes. A comprehensive review of meta-analyses of observational and randomized trial studies suggests that MediD intervention is associated with protective effects in case of colorectal, liver and pancreatic cancer. Null results are found with bladder, endometrial and ovarian cancer, while the effect of MediD in case of neck/head, respiratory, oesophageal, prostate, gastric and breast cancers was found to be controversial (Dinu et al. 2018).

A vegan dietary style, which mostly consists of whole grains, legumes, nuts, seeds, vegetables, fruit and exclusion of all animal products, is one of the popular dietary patterns of cancer survivors. A meta-analysis comprising of 4 prospective cohort studies and 24 cross-sectional studies revealed that vegan diets were found to be linked with decrease in total cancer incidence risk by 15% (Dinu et al. 2017). The outcome of series of small randomized clinical trials has demonstrated the effect of low-fat vegan diet in reducing the prostate-specific antigen levels (Ornish et al. 2005), positively influencing prostate gene expression (Ornish et al. 2008) and improving overall health and reducing mortality (Le and Sabaté 2014). A study of 2246 vegans from the Oxford Vegetarian Study and EPIC-Oxford showed 19% reduction in the risk of developing any cancer; however, the finding was found nonsignificant when the data were correlated with BMI (Key et al. 2014). Under similar settings, three Adventists cohort study analysis demonstrated 14% reduction in the risk of all cancer, but the study results suggested more benefit of vegan diet to males than females (Le and Sabaté 2014). Data analyses of the Oxford Vegetarian

Study and EPIC-Oxford settings also pointed out that the prevalence of prostate cancer was extensively reduced by 35% in the vegan cohort as compared with nonvegetarians, but here also the benefits were more in favour with white race and the outcome was nonsignificant in case of blacks (Tantamango-Bartley et al. 2016). The Oxford Vegetarian and EPIC-Oxford study also revealed that a vegan diet follow-up range of 4.14–7.8 years was not connected with reduction in cancer risk of colorectal, breast, respiratory tract, gastrointestinal or urinary tract cancers (Key et al. 2014; Le and Sabaté 2014; Orlich et al. 2015; Penniecook-Sawyers et al. 2016; Tantamango-Bartley et al. 2016). The best part of the Vegan dietary pattern is that it qualifies majority of diet recommendations advocated by ACS and AICR and therefore is a scope and opportunity for fine-tuning concordance of this dietary style by conducting more clinical studies under defined set of cancer conditions.

The alkaline dietary pattern was introduced with a logic and great enthusiasm of making it as adjuvant cancer therapy. The philosophy of the alkaline diet (fruits and vegetables of alkaline nature) is based on fact that acidic extracellular environment is more favourable for cancer cell proliferation, invasion and metastasis. The alkaline diet was introduced with a hope of creating an alkaline microenvironment that may help in cancer prevention by discouraging cancer cell proliferation, invasion and metastasis. There are very few studies carried out in support with alkaline diet as cancer prevention pattern. For example, in recent epidemiological study, it has been observed that diet-dependent acid load is linked with risk of breast cancer and the inclusion of fruits and vegetables and moderate consumption of meat may be protective against hormone receptor-negative breast cancer. The study lacks the mechanistic explanation about alkaline diet (Park et al. 2019). In another human study, alkaline diet intervention enhanced the effect of EGFR-TKI treatment in non-small cell lung cancer patients with EGFR mutations (Hamaguchi et al. 2017). The study was devoid of control group; therefore, interpretation and outcome of the study have limited significance. A systematic review of literature comprising of many cross-sectional studies has concluded that there is lack of convincing clinical evidence linking the alkaline diet role in the prevention of cancer, and hence, the alkaline diet promotion for treatment or prevention of cancer may not be acceptable (Fenton and Huang 2016); however, for cancer survivors, with dietary recommendations the alkaline diet is greatly consistent.

Developing specific amino acid-deficient diets is one more approach that has been studied for their abilities for inhibiting the several tumour models' growth. For instance, normal cells are capable of synthesizing sufficient amount of methionine for their growth and development using vitamin B12 and 5-methyltetrahydrofolate and homocysteine. Nevertheless, several types of cancer cells are dependent on exogenous supply of methionine for the growth and survival. Methionine sensitivity of cancer cells is being explored as an opportunity to starve the cancer cells by offering methionine-restricted (MR) diet, which will arrest the growth of cancer cells (Chaturvedi et al. 2018). The studies of preclinical animal model have shown impressive anticancer activity and strongly support the strategy of developing methionine-restricted diet as therapeutic approach (Gao et al. 2019; Jeon et al. 2016; Hens et al. 2016). For glioma or melanoma, an MR diet's phase II clinical trial in combination with cystemustine treatment revealed that the plasma level of methionine was significantly reduced by 58% on the first day of the MR diet with a 1.8 months of average disease-free survival and 4.6 months of median survival (Hoffman 2019). Another amino acid having potential to starve the cancer cells is serine and glycine. These amino acids are biosynthetically linked and act as essential precursors for the biosynthesis of nucleic acids, proteins and lipids that are essential for growth of cancer cells. Moreover, hyperactivation of the serine/glycine biosynthetic pathway is strongly associated with oncogenesis, and hence, exploiting the addiction of tumour with a glycine and serine-free diet might provide a novel translational opportunity for dietary intervention, drug development and biomarker identification of human cancers (Amelio et al. 2017). Preclinical animal model studies have revealed glycine/serine's dietary restriction extending the survival of mouse with lymphoma and intestinal cancer (Maddocks et al. 2017). Mechanistic clinical evidence of methionine, serine and glycine free diets might pave a way for clinical benefits of these amino acid-free dietary patterns.

Both the macrobiotic and Palaeolithic diets have philosophical, cultural and ethnic grounds and the negative concerns of these dietary patterns are the pseudo-scientific rationales of their cancer-protective properties, poor evidence about improved cancer outcomes and perhaps the exclusion of cancer-preventing and health-promoting food ingredients. Except few studies, there are no evolving clinical investigations towards the cancer-ameliorating effects of macrobiotic and Palaeolithic dietary styles. Selection of evidence-based dietary pattern can have a substantial impact on mortality, survival, incidence and cancer risk reduction (Bamia 2018).

Conclusion

The dietary patterns preferred by the contemporary cancer survivors are not fully in tune with the WCRF/AICR and ACS dietary recommendations owing to inadequate research on these popular diets' impact on cancer incidence, mortality and overall improvement in quality of life of cancer survivors. Perhaps such lacunae may create confusion in clinicians leading to not providing concrete advice to their cancer patients on the implementation of these dietary patterns for the management of cancer. Therefore, there is a research opportunity to validate the individual dietary pattern in concert with mechanistic clinical evidence so as to strongly advocate the benefits of these popular dietary styles.

References

Amelio I, Melino G, Frezza C. Exploiting tumour addiction with a serine and glycine-free diet. Cell Death Differ. 2017;24(8):1311–3.

Baden MY, Liu G, Satija A, Li Y, Sun Q, Fung TT, et al. Changes in plant-based diet quality and total and cause-specific mortality. Circulation. 2019;140(12):979–91. https://doi.org/10.1161/CIRCULATIONAHA.119.041014.

Bamia C. Dietary patterns in association to cancer incidence and survival: concept, current evidence, and suggestions for future research. Eur J Clin Nutr. 2018;72(6):818–25.

Chaturvedi S, Hoffman RM, Bertino JR. Exploiting methionine restriction for cancer treatment. Biochem Pharmacol. 2018;154:170–3.

Chikara S, Nagaprashantha LD, Singhal J, Horne D, Awasthi S, Singhal SS. Oxidative stress and dietary phytochemicals: role in cancer chemoprevention and treatment. Cancer Lett. 2018;413:122–34.

Cohen CW, Fontaine KR, Arend RC, Gower BA. A ketogenic diet is acceptable in women with ovarian and endometrial cancer and has no adverse effects on blood lipids: a randomized, controlled trial. Nutr Cancer. 2020;72(4):584–94. https://doi.org/10.1080/01635581.2019.1645864.

Dinu M, Abbate R, Gensini GF, et al. Vegetarian, vegan diets and multiple health outcomes: a systematic review with meta-analysis of observational studies. Crit Rev Food Sci Nutr. 2017;57:3640–9.

Dinu M, Pagliai G, Casini A, Sofi F. Mediterranean diet and multiple health outcomes: an umbrella review of meta-analyses of observational studies and randomised trials. Eur J Clin Nutr. 2018;72(1):30–43.

Fenton TR, Huang T. Systematic review of the association between dietary acid load, alkaline water and cancer. BMJ Open. 2016;6:e010438. https://doi.org/10.1136/bmjopen-2015-010438.

Fortes C, Forastiere F, Farchi S, et al. The protective effect of the Mediterranean diet on lung cancer. Nutr Cancer. 2003;46(1):30–7.

Gao X, Sanderson SM, Dai Z, Reid MA, Cooper DE, Lu M, et al. Dietary methionine influences therapy in mouse cancer models and alters human metabolism. Nature. 2019;572(7769):397–401.

Giraldi L, Panic N, Cadoni G, Boccia S, Leoncini E. Association between Mediterranean diet and head and neck cancer: results of a large case-control study in Italy. Eur J Cancer Prev. 2017;26(5):418–23.

Hamaguchi R, Okamoto T, Sato M, Hasegawa M, Wada H. Effects of an alkaline diet on EGFR-TKI therapy in EGFR mutation-positive NSCLC. Anticancer Res. 2017;37(9):5141–5.

Hens JR, Sinha I, Perodin F, Cooper T, Sinha R, Plummer J, Perrone CE, Orentreich D. Methionine-restricted diet inhibits growth of MCF10AT1-derived mammary tumors by increasing cell cycle inhibitors in athymic nude mice. BMC Cancer. 2016;16:349.

Hoffman RM. Clinical studies of methionine-restricted diets for cancer patients. Methods Mol Biol. 2019;1866:95–105.

Jalilpiran Y, Dianatinasab M, Zeighami S, Bahmanpour S, Ghiasvand R, Mohajeri SAR, Faghih S. Western dietary pattern, but not Mediterranean dietary pattern, increases the risk of prostate cancer. Nutr Cancer. 2018;70(6):851–9. https://doi.org/10.1080/01635581.2018.1490779.

Jeon H, Kim JH, Lee E, Jang YJ, Son JE, Kwon JY, Lim TG, Kim S, Park JH, Kim JE, Lee KW. Methionine deprivation suppresses triple-negative breast cancer metastasis in vitro and in vivo. Oncotarget. 2016;7(41):67223–34.

Key TJ, Appleby PN, Crowe FL, Bradbury KE, Schmidt JA, Travis RC. Cancer in British vegetarians: updated analyses of 4998 incident cancers in a cohort of 32,491 meat eaters, 8612 fish eaters, 18,298 vegetarians, and 2246 vegans. Am J Clin Nutr. 2014;100(Suppl 1):378S–85S.

Khodabakhshi A, Akbari ME, Mirzaei HR, Mehrad-Majd H, Kalamian M, Davoodi SH. Feasibility, safety, and beneficial effects of MCT-based ketogenic diet for breast cancer treatment: a randomized controlled trial study. Nutr Cancer. 2020;72(4):627–34. https://doi.org/10.1080/01635581.2019.1650942.

Klement RJ. The emerging role of ketogenic diets in cancer treatment. Curr Opin Clin Nutr Metab Care. 2019;22(2):129–34.

Klement RJ, et al. Impact of a ketogenic diet intervention during radiotherapy on body composition: III-final results of the KETOCOMP study for breast cancer patients. Breast Cancer Res. 2020;22 (1):94.

Le LT, Sabaté J. Beyond meatless, the health effects of vegan diets: findings from the Adventist cohorts. Nutrients. 2014;6(6):2131–47.

Liese AD, Krebs-Smith SM, Subar AF, et al. The Dietary Patterns Methods Project: synthesis of findings across cohorts and relevance to dietary guidance. J Nutr. 2015;145:393–402.

Maddocks ODK, Athineos D, Cheung EC, Lee P, Zhang T, van den Broek NJF, et al. Modulating the therapeutic response of tumours to dietary serine and glycine starvation. Nature. 2017;544 (7650):372–6.

Mayne ST, Playdon MC, Rock CL. Diet, nutrition, and cancer: past, present and future. Nat Rev Clin Oncol. 2016;13:504–15.

Mentella MC, Scaldaferri F, Ricci C, Gasbarrini A, Miggiano GAD. Cancer and Mediterranean diet: a review. Nutrients. 2019;11(9). pii: E2059. https://doi.org/10.3390/nu11092059.

Na BG, Han SS, Cho YA, Wie GA, Kim JY, Lee JM. Nutritional status of patients with cancer: a prospective cohort study of 1,588 hospitalized patients. Nutr Cancer. 2018;70(8):1228–36.

Orlich MJ, Singh PN, Sabaté J, et al. Vegetarian dietary patterns and the risk of colorectal cancers. JAMA Intern Med. 2015;175(5):767–76.

Ornish D, Weidner G, Fair WR, et al. Intensive lifestyle changes may affect the progression of prostate cancer. J Urol. 2005;174:1065–9; discussion 9–70.

Ornish D, Magbanua MJ, Weidner G, et al. Changes in prostate gene expression in men undergoing an intensive nutrition and lifestyle intervention. Proc Natl Acad Sci U S A. 2008;105:8369–74.

Park YM, Steck SE, Fung TT, Merchant AT, Elizabeth Hodgson M, Keller JA, Sandler DP. Higher diet-dependent acid load is associated with risk of breast cancer: findings from the sister study. Int J Cancer. 2019;144(8):1834–43.

Patel A, Pathak Y, Patel J, et al. Role of nutritional factors in pathogenesis of cancer. Food Quality Safety. 2018;2(1):27–36.

Penniecook-Sawyers JA, Jaceldo-Siegl K, Fan J, et al. Vegetarian dietary patterns and the risk of breast cancer in a low-risk population. Br J Nutr. 2016;115(10):1790–7.

Plotti F, Terranova C, Luvero D, et al. Diet and chemotherapy: the effects of fasting and ketogenic diet on cancer treatment. Chemotherapy. 2020;65(3–4):77–84.

Römer M, Dörfler J, Huebner J. The use of ketogenic diets in cancer patients: a systematic review. Clin Exp Med. 2021. https://doi.org/10.1007/s10238-021-00710-2.

Russo GI, Solinas T, Urzì D, Privitera S, Campisi D, Cocci A, Carini M, Madonia M, Cimino S, Morgia G. Adherence to Mediterranean diet and prostate cancer risk in Sicily: population-based case-control study. Int J Impot Res. 2019;31(4):269–75.

Salvatore Benito A, Valero Zanuy MÁ, Alarza Cano M, Ruiz Alonso A, Alda Bravo I, Rogero Blanco E, et al. Adherence to Mediterranean diet: a comparison of patients with head and neck cancer and healthy population. Endocrinol Diabetes Nutr. 2019;66(7):417–24.

Schulpen M, van den Brandt PA. Adherence to the Mediterranean diet and risk of lung cancer in the Netherlands Cohort Study. Br J Nutr. 2018;119(6):674–84.

Schulpen M, van den Brandt PA. Mediterranean diet adherence and risk of colorectal cancer: the prospective Netherlands Cohort Study. Eur J Epidemiol. 2020;35(1):25–35. https://doi.org/10.1007/s10654-019-00549-8.

Schwartz KA, Noel M, Nikolai M, Chang HT. Investigating the ketogenic diet as treatment for primary aggressive brain cancer: challenges and lessons learned. Front Nutr. 2018;5:11.

Serra-Majem L, Román-Viñas B, Sanchez-Villegas A, Guasch-Ferré M, Corella D, La Vecchia C. Benefits of the Mediterranean diet: epidemiological and molecular aspects. Mol Aspects Med. 2019;67:1–55.

Shaikh AA, Braakhuis AJ, Bishop KS. The Mediterranean diet and breast cancer: a personalised approach. Healthcare (Basel). 2019;7(3). pii: E104.

Tantamango-Bartley Y, Knutsen SF, Knutsen R, et al. Are strict vegetarians protected against prostate cancer? Am J Clin Nutr. 2016;103(1):153–60.

Turati F, Bravi F, Polesel J, Bosetti C, Negri E, Garavello W, et al. Adherence to the Mediterranean diet and nasopharyngeal cancer risk in Italy. Cancer Causes Control. 2017;28(2):89–95.

US Department of Agriculture, Dietary Guidelines Advisory Committee. Scientific Report of the 2015 Dietary Guidelines Advisory Committee. 2015. health.gov/dietaryguidelines/2015-scientific-report/. Accessed 15 Nov 2020.

US Department of Health and Human Services, US Department of Agriculture. 2015-2020 Dietary guidelines for Americans. US Department of Health and Human Services; 2015.

van der Louw EJTM, Olieman JF, van den Bemt PMLA, Bromberg JEC, Oomen-de Hoop E, et al. Ketogenic diet treatment as adjuvant to standard treatment of glioblastoma multiforme: a feasibility and safety study. Ther Adv Med Oncol. 2019;11:1758835919853958. https://doi.org/10.1177/1758835919853958.

Wang Y, et al. Does a ketogenic diet as an adjuvant therapy for drug treatment enhance chemotherapy sensitivity and reduce target lesions in patients with locally recurrent or metastatic Her-2-negative breast cancer? Study protocol for a randomized controlled trial. Trials. 2020;21(1):487.

Weber DD, Aminzadeh-Gohari S, Tulipan J, Catalano L, Feichtinger RG, Kofler B. Ketogenic diet in the treatment of cancer—where do we stand? Mol Metab. 2019. pii: S2212-8778(19)30427-2.

World Cancer Research Fund/American Institute for Cancer Research. Continues Update Project Expert Report. Recommendations and public health and policy implications. 2018. dietandcancerreport.org. Accessed 24 Sept 2020.

Zahra A, Fath MA, Opat E, et al. Consuming a ketogenic diet while receiving radiation and chemotherapy for locally advanced lung and pancreatic cancer: the University of Iowa experience of two phase I clinical trials. Radiat Res. 2017;187:743–54.

Zhang FF, Cudhea F, Shan Z, et al. Preventable cancer burden associated with poor diet in the United States. JNCI Cancer Spectr. 2019a;3(2):pkz034.

Zhang X, Tang T, Pang L, Sharma SV, Li R, Nyitray AG, Edwards BJ. Malnutrition and overall survival in older adults with cancer: a systematic review and meta-analysis. J Geriatr Oncol. 2019b;10(6):874–83.

Zick SM, Snyder D, Abrams DI. Pros and cons of dietary strategies popular among cancer patients. Oncology (Williston Park). 2018;32(11):542–7.

Chapter 3
Frequently Asked Diet- and Nutrition-Related Questions by Cancer Survivors

"Difficult roads can lead to beautiful destinations."………….Kia Wynn, oral cancer survivor

Abstract Intake of healthy diet and nutrition is a critical part for the management of any human ailments including cancer. It has been well-established fact that the treatment alone may not be possible to cure the disease, but the proactive role played by our immune system towards the drugs is equally important. Undoubtedly, diet and nutrition play a key role in maintaining our immune system in a functional mode. Series of diet- and nutrition-related questions raised by cancer survivors are still waiting to be addressed scientifically. Surprisingly, many of the world's renowned healthcare agencies and hospitals are either reluctant or provided mixed and consistently inconsistent information towards the questions raised by cancer patients. This chapter describes the frequently asked questions by cancer survivors along with possible answers. The present state-of-the-art situation warrants more focused research for addressing a major question of cancer survivors as to what is the optimal diet requirement during and after cancer treatments. The entire discourse might inspire the scientific community to undertake research activity for resolving the questions being asked by cancer patients.

Keywords Diet · Nutrition · Cancer survivors · FAQ · Cancer organizations/ agencies

Abbreviations

ACS	American Cancer Society
AICR	American Institute for Cancer Research
CDC	The Centres for Disease Control and Prevention
FDA	Food and Drug Administration
NCCN	National Comprehensive Cancer Network
NCI	National Cancer Institute
USDA	The United States Department of Agriculture

R. N. Gacche, *Dietary Research and Cancer*,
https://doi.org/10.1007/978-981-16-6050-4_3

WCRF World Cancer Research Fund
WHO World Health Organization

Introduction

The Overview of Frequently Asked Questions by the Cancer Survivors

Majority of the most frequently asked questions by the cancer survivors in relation to diet and nutrition pertain to the kind of personalized clinical symptoms, enquiry about the inclusion of dietary supplements and preferences of dietary patterns and foods. The most commonly observed side effect of chemotherapy in cancer survivors includes nausea and vomiting, fatigue, loss of appetite and weight loss, anaemia, dehydration, diarrhoea, delirium, hair loss, infertility and sexual dysfunction, sleep disorders, oedema, mouth problems (ulcers and difficulty in swallowing), difficulty in breathing, skin problems, constipation and pain. (National Cancer Institute (NCI): Chemotherapy and You 2011; Patient Page: CA Cancer J Clin 2003, 2012). Among the bunch of side effects, the cancer patients enquire about the food and nutrition requirements that may ameliorate the aforesaid side effects; for example, which foods will be helpful, in case of dry mouth and swallowing problems? Are there foods that will help to restore lost appetite coupled with nausea? What to eat to reduce fatigue? Is there any special diet to help improve anaemia? Is there need to make any changes to dietary plan once an individual is diagnosed with cancer? Are there certain foods that should be prioritized or should be avoided? Is there any specific diet that one should follow during chemotherapy or radiation? How safe it is to take vitamin, antioxidants or dietary supplements during and after chemotherapy or radiation therapy? Cancer patients even write to several healthcare-taking agencies and hospitals and register their questions. For example, series of questions like "How Will Good Nutrition Help Me During Cancer Treatment?"; "How Does My Dietitian Assist in My Care?"; "Does High Sugar Intake Increase My Risk of Cancer?" "Does Aspartame Cause Cancer?"; "Will I Lower My Risk of Cancer if I Eat More Fruits and Vegetables?"; "Are Organic Foods Better Than Conventional Foods for Reducing Cancer Risk?"; "What Is a Plant-Based Diet?"; "What Are Antioxidants?"; and "What Are Phytochemicals?" have been posted at the home site of Mass General Brigham, Newton-Wellesley Hospital, Newton, MA, USA (https://www.nwh.org/mass-general-cancer-center/survivorship-care/cancer-nutrition-services/faqs).

Few more interesting diet–cancer-related questions like "Does sugar feed cancer?"; "Is soy safe for cancer patients?"; "Should I buy organic food?"; "Is fasting safe during cancer treatment?"; and "Do superfoods have exceptional powers for fighting illness?" are posted at the home page of a reputed cancer institute, Dana Farber Cancer Institute, Boston, CA, USA (https://www.dana-farber.org/for-patients-and-families/care-and-treatment/support-services-and-amenities/nutrition-services/faqs/). An interesting set of questions like "Can fasting or calorie restriction help my body fight cancer?"; "Could it also help cancer treatment be more effective?"; "Does sugar feed cancer, and if so, should I completely avoid it?"; "Should I restrict my carbohydrate intake to help my body fight cancer?"; "Do all animal foods increase the risk for cancer? What factors within animal food are problematic?"; "Should I include supplements and/or herbal medicine in my fight against cancer?"; and "What can I do to help minimize side effects during treatment?" are also found at home page of Osher Centre for Integrative Medicine, University of California San Francisco, USA (https://osher.ucsf.edu/patient-care/integrative-medicine-resources/cancer-and-nutrition/faq).

Common questions asked by cancer patients about diet, activity and cancer risk have been posted American Cancer Society, and the questions mostly related to diet includes "What are antioxidants, and what do they have to do with cancer?"; "Does drinking coffee affect cancer risk?"; "Does eating a gluten-free diet help reduce cancer risk?"; "Do anti-inflammatory diets reduce cancer risk?"; "Why are foods irradiated, and can these foods increase cancer risk?"; "Can periods of limiting food intake to juices remove toxins and help protect against cancer?"; "Do non-nutritive sweeteners/sugar substitutes cause cancer?"; "Are foods labelled 'organic' more effective in lowering cancer risk?"; "Do pesticides in foods cause cancer?"; "Can soy-based foods reduce cancer risk?"; "Does sugar increase cancer risk?"; and "Do vegetarian diets reduce cancer risk?" (https://www.cancer.org/healthy/eat-healthy-get-active/acs-guidelines-nutrition-physical-activity-cancer-prevention/common-questions.html).

It is clear from the variety of questions being asked by cancer patients across world that there is lack of information regarding the availability of consistent and evidence-based diet-related information to cancer in general and with types of cancer in particular. This can be even evidenced from a study, which was designed to assess the dietary recommendations for cancer patients during treatment and overall survival. The study evaluated the websites of all National Comprehensive Cancer Network (NCCN) member institutions, which are available at www.nccn.org, and individual sites were reviewed for recommendations about nutritional recommendations mostly covering the diet type, meal frequency, macronutrient content and other specific diet-related recommendations. Of the 21 NCCN member institutions, only 4 sites (19%) provided nutrition-related guidelines. Half of the sites recommended a low-fat, high-carbohydrate diet (5:1) and carbohydrate-to-fat (7:1) food types, and half suggested maintaining weight during cancer treatment with a recommendation of intake of carbohydrate to fat (1:1). One third of all NCCN sites available had established links to nine external diet-related webs. In brief, there is lack of consistent online dietary recommendations for cancer patients, which perhaps they may implement during and after cancer treatment. This clearly advocates the need of conducting future research to develop evidence-based cancer-specific guidelines, which may be referred by oncologists and patients as well (Champ et al. 2013).

Even a recent systematic review of prospective cohort studies was designed to investigate the associations of postdiagnostic diet with prognostic outcomes in cancer patients. A total of 29 prospective studies of different cancers were considered. The study concluded that intake of low-fat diet, regular consumption of fibre containing vegetables, healthy quality diet and high-quality protein consumption were found beneficial to cancer patient, while Western diet (WD) along with high intake of saturated fats was associated with a higher risk of mortality. The study suggests to avoid the WD pattern and also recommended that none of the food categories including meat and dairy products should be eliminated in diets of cancer patients. The study report advocates more precise and focused evidence-based studies for the development of consistent dietary guidelines for cancer patients (Rinninella et al. 2020).

The Common Questions Asked by the Cancer Survivors and Answer by the Experts of ACS

ACS happens to be a premier organization working for welfare of cancer patients. ACS routinely publish the "Diet and Physical Activity Guideline", which serves as a baseline foundation for awareness, communication, making policies and developing community strategies of diet and physical activity patterns for the Americans. The frame of guidelines is formulated by a panel of national experts from cancer research, epidemiology, prevention, policy and public health, and it focuses on the most recent scientific evidences, which are associated with dietary research, physical activity patterns and related cancer risk. The recently published ACS guidelines 2020 towards diet and physical activity for cancer prevention have provided an elaborative section on common questions asked by the cancer survivors and also provided rational answers. As far as dietary questions are concerned, they are mostly related to antioxidants, the safety of genetically modified foods, impact of coffee on cancer risk, gluten-free diet and cancer risk, irradiated foods and risk of cancer, fruit juices and protection against cancer, intake of anti-inflammatory foods and cancer risk, relation of sweeteners and sugar with causation of cancer, pesticides residues in foods and onset of cancer, link of soya and soya-derived foods with cancer risk and association of vegetarian diet with cancer risk (Rock et al. 2020). The expert panel based on the circumstantial literature have tried to provide judicious answers to every question, for example, according to the experts.

It is advised to consume antioxidants through dietary sources, rather than relying on antioxidant supplements (WCRF/AICR, Continuous Update Project. The Third Expert Report 2018). The experts opined that in the present state of the art there is no evidence reporting the harmful effect of genetically engineered food ingredients either on human health or either negatively or positively influence the risk of cancer. While answering the impact of coffee, the authors say that consumption of coffee is likely to lower the risk of endometrial, liver, pharynx, mouth, larynx, skin cancer, etc., and possibly may prevent malignant melanoma in women. There is also a suggestion that there can be advantages of drinking coffee at moderate temperature as hot coffee may be a risk factor for oesophageal cancer.

As per the ASC expert opinion, there is very scanty evidence relating to intake of gluten and risk of gastrointestinal cancers. Nevertheless, there is no evidence that relates the intake of gluten-free diet lowering the risk of risk; on the contrary, several studies have demonstrated that consumption of gluten-containing whole grains may lower the risk of colon cancer. While answering to the effect of irradiated food and cancer risk, the experts clear the confusion and stated that the process of food irradiation does not impart radioactivity to foods, does not affect the nutritional quality or even has no effect on texture, taste and appearance of irradiated foods (IFs). The US FDA (Food and Drug Administration) has conducted experiments and evaluated the safety of IFs over period of more than 30 years and confirmed the safety of irradiated foods. The WHO, USDA and the CDC have also endorsed the safety issues of IFs. Therefore, in the mainstream of contemporary research there is no scientific evidence that links the consumption of IFs with causation of cancer or there is no evidence that attributes the harmful effects of IFs on human health (US Food and Drug Administration 2018; Ravindran and Jaiswal 2019). While making comments on fruit juices and protection against cancer, the expert panel has described that there exist no scientific data in support of the claims that daily exclusive consumption of fruit juices lowers the risk of cancer or otherwise provides any apparent health benefits. Nevertheless, the popularity of juices as cleansers or detoxifying agents is also not supported by scientific literature; however, vegetable juices may help to increase micronutrient levels in the body (Butalla et al. 2012). On the contrary, a dietary intake limited to juice may not provide adequate important nutrients and, in selected cases, may also contain high concentration of some toxic substances that may cause damage to kidney and create adverse health problems (Hu 2019; Lien 2013; Tavernise 2016).

The term "organic food" has become a buzz word, and many consumers have opinion that the intake organic foods may be helpful on health grounds; however, the expert says that there are little scientific data, which advocate that organic stuff possess higher levels of nutrient as compared to traditionally grown agricultural produce. A very scanty research data are available, which links the consumption of organic food with risk of cancer; on the other hand, a recent investigation observed the negative concern of consuming organic food and risk of non-Hodgkin lymphoma; however, such outcomes need to be verified under different settings (Baudry et al. 2018). Answering the frequently asked important question, whether sugar intake increases the cancer risk? The expert opinion and the circumstantial literature outline the fact that different types of sugars in foods and other products like beverages significantly contribute to caloric additions and thereby promote obesity. Moreover, excess consumption of sugar can indirectly increase the risk of cancer. The preclinical studies have already demonstrated that cancer cells have higher rates of glucose metabolism as compared to normal cells. There exists a scientific evidence, which states that a dietary pattern enriched with added sugars may adversely affect the levels of insulin and other sugar-related hormones involved in sugar homeostasis and may increase the risk of certain types of cancers (Perez-Hernandez et al. 2014).

The Possible Answers to Cancer Survivors by Other Organizations/Healthcare Agencies

Unfortunately, not all patients (especially illiterate patients) are not able to access the available dietary recommendations or guidelines and majority of the patients make an online search for this information. Before describing the possible answers to the questions asked by the cancer survivors, it is imperative to understand certain facts and ground realities. For example, it is very common clinical observations that the loss of appetite is the most common side effect for cancer patients who are undergoing cancer treatment, and even the degree of appetite loss varies from patient to patient and even with type of cancer. Second most important observation is that the cancer patient complains about the loss or change of taste of food. The symptoms like nausea and other adversities discourage the cancer patient from taking food, although with progress of cancer the eating also becomes progressively difficult for some cancer patients. However, besides these adversities, good nutrition and diet and exercise remain significant hope for maintaining the body weight, quality of life and strength of cancer patient during the treatment duration and even after the drug holidays. There can be one general suggestion to cancer patients regarding finding possible answers towards diet-related questions. As suggested by National Cancer Institute (cancer.gov), "Patients with cancer should receive evidence-based, individualized recommendations from a nutrition expert, based on a number of factors, including their type of cancer, stage of disease, planned treatment, nutrition history, weight history, and lab values" (https://www.cancer.gov/about-cancer/treatment/research/nutrition-during-after-treatment). One more important thing that needs to be understood by cancer patients is that the great strides of research in molecular biology of cancer have unravelled several secretes of cancer progression, and it is also suggested that it may not be possible to apply global dietary recommendations to patients with different types of cancer having progression at various stages (Bose et al. 2020). For example, patients of locally advanced head and neck cancers can have great difficulty in eating during treatment duration and are often suggested to eat compatible diet, which can keep weight at normal level (Crowder et al. 2018). On the other hand, the patients with localized type of cancers, such as prostate cancer or breast cancer, may eat per their will and wish; however, this liberty of whatever they want may adversely affect their health; for example, if these patients are likely to become long-term survivors, they may gain weight and may increase the risk of cancer relapse (Engin 2017; Yuan et al. 2020).

Although concrete evidence-based scientific answers to the above questions are not available, however, several National and International cancer-related organizations/agencies like AICR (www.aicr.org), CSC (www.cancersupportcommunity.

org), NCI (www.cancer.gov) and many cohort and observational studies have provided the experience-based practical answers to majority of the frequently asked questions by cancer survivors. NCI's informative resource book "Eating Hints: Before, during, and after Cancer Treatment" (www.cancer.gov/publications/patient-education/eating-hints) is a compilation of hints about common types of eating problems associated with cancer patients and the ways to manage them using different food types along with variety of recipes for cancer survivors. For example, while describing the ways for managing nausea and vomiting, NCI has recommended the antinausea medicine along with fruit juices, ginger ale, tea and/or sports drinks and also suggested to avoid greasy, fried, sweet or spicy foods.

The recent systematic literature analyses of randomized controlled trials also suggest that ginger may ameliorate the nausea in the acute phase of chemotherapy in patients with breast cancer (Totmaj et al. 2019). For management of diarrhoea and constipation, AICR has suggested drinking of more healthy beverages (water, prune juice, warm juices, decaffeinated teas and hot lemonade) to keep digestive system moving, increase intake of high-fibre foods such as whole grains, fresh and cooked vegetables, fresh and dried fruits, and foods containing peels, nuts and seeds. There is also advice of using stool softeners and gentle, non-habit-forming laxatives, etc. ("HEAL Well: A Cancer Nutrition Guide") by AICR (www.aicr.org/assets/docs/pdf/education/heal-well-guide.pdf). A recent qualitative review analysis suggests that increased protein intake and choosing the dietary patterns that reduce inflammation (mostly the Mediterranean diet and other plant-based diets) and supplementation with ginger, ginseng or probiotics may improve cancer survivors' energy levels and may help to reduce the fatigue level (Julia et al. 2019). The judicious selection of dietary patterns that reduces inflammation, inclusion of plant-based diets and supplementation of ginger, ginseng or probiotics has been qualitatively reviewed in relation to the management of cancer-related fatigue (Inglis et al. 2019).

Of the many healthcare agencies and hospitals addressing the answers to the questions frequently asked by cancer patients, Mass General Brigham, Newton-Wellesley Hospital, Newton, (MA), USA, has posted rational and easy to understand answers to cancer patients. For example, while answering a simple question as to "How Will Good Nutrition Help Me During Cancer Treatment?" the simple answer given is "Eating well and maintaining good nutrition before, during, and after cancer treatment can help you maintain strength and energy while you go through treatment. A healthy diet also can help you better deal with any side effects you might experience from treatment". Other questions as "Does High Sugar Intake Increase My Risk of Cancer?", perhaps this is a misconception and fear in the minds of many cancer patients, the answer provided is "Sugar intake isn't directly linked to an increased cancer risk". However, diets high in sugar often contain lower amounts of nutritious foods such as fruit, vegetables, whole grains and beans. High-sugar diets also increase your risk of being overweight and having higher levels of insulin production, which can increase your risk for diabetes. Dana Farber Cancer Institute, Boston, (CA), USA, has also provided very elaborative scientific explanations to most important questions like "Is fasting safe during cancer treatment?", and the answer provided is "Fasting and calorie restriction (CR) have been areas of interest in both prevention of cancer, as well as reducing side effects from cancer treatments. There are limited studies looking at the impact of fasting/CR and incidence of cancer, and caution must be used in interpreting these results given the studies vary in type and quality. Fasting during the window around chemotherapy has gained attention as a possible way to reduce common side effects of treatment. The thinking is that when nutrient intake is significantly reduced or withheld, healthy cells respond to this stress and can slow growth and shift towards maintenance, whereas cancer cells cannot. This may cause cancer cells, but not healthy cells, to experience the damaging effects of treatment. While a few small trials show some benefit in reducing side effects of treatment, larger-scale trials are needed to further assess the safety and effectiveness of this type of dietary intervention. The need for more data is a particularly important considering many patients already find it challenging to meet nutrition needs during cancer treatment". They have also provided a logical answer to an important question like "Do superfoods have exceptional powers for fighting illness?" and answer provided is "The term superfoods has become synonymous with ultra-healthy, nutrient-packed, plant-based foods and supplements". They are advertised as having exceptional powers for fighting or preventing common health conditions, like cancer or diabetes, and promoting everything from energy and vitality to hormonal balance. New superfoods are constantly being introduced, adding to the saturated health–food marketplace and contributing to consumer confusion. Many patients ask us, "What are superfoods?" and "Do we need to eat these unusual foods in order to be healthy?" Many superfoods are simply colourful, phytonutrient-rich fruits, vegetables, herbs, nuts, seeds or legumes from around the globe. Superfoods, just like regular plant foods, provide fibre, vitamins, minerals, antioxidants, anti-inflammatory and immune-supportive nutrients. While "superfoods" are often marketed as a panacea or magical elixir that can improve upon or enhance one's ability to prevent or reduce cancer, the scientific evidence is sparse. It is important to note that current studies are preliminary. They are conducted on cells or animals in the laboratory, not humans. Even though they may not always live up to their hype, superfoods can be a part of a well-balanced diet and offer new and interesting options for delicious plant-based meals. What we do know: a balanced diet plentiful in plant-based foods, such as fruits, vegetables, nuts, seeds and whole grains, along with healthy fats, and lean proteins like fish, peas, lentils and other lean proteins, in combination with regular physical activity and weight management, may help reduce the risk for developing certain cancers and may help promote survivorship.

Conclusion

More comprehensive evidence-based clinical studies rationalized on the questions of the cancer survivors need to be carried out for giving satisfactory answers to diet-related questions of the cancer patients. Such endeavours may help to remove the diet-related unscientific beliefs, illusions and fear from the minds of cancer patients.

Equally, the physicians, oncologists, cancer care providers and other stakeholders need to be proactive in integrating the dietary interventions in clinical practice rather than showing negligence towards this important issue of dietary interventions in the integrated management of cancer. Most important, the cancer patients should always refer the suggestions, recommendations and guidelines of dedicated cancer-related agencies like ACS, WCRF, NCCN, NCI and AICR rather than getting victimized to commercial cosmetic diet-related advertisements.

References

Baudry J, Assmann KE, Touvier M, et al. Association of frequency of organic food consumption with cancer risk: findings from the NutriNet-Sante prospective cohort study. JAMA Intern Med. 2018;178:1597–606.

Bose S, Allen AE, Locasale JW. The molecular link from diet to cancer cell metabolism. Mol Cell. 2020;78(6):1034–44.

Butalla AC, Crane TE, Patil B, et al. Effects of a carrot juice intervention on plasma carotenoids, oxidative stress, and inflammation in overweight breast cancer survivors. Nutr Cancer. 2012;64:331–41.

Champ CE, Mishra MV, Showalter TN, Ohri N, Dicker AP, Simone NL. Dietary recommendations during and after cancer treatment: consistently inconsistent? Nutr Cancer. 2013;65(3):430–9.

Chemotherapy and You: Support for People with Cancer, a booklet from the National Cancer Institute. NIH Publication No. 11-7156. 2011. www.cancer.gov/publications/patient-education/chemo-and-you.

Crowder SL, Douglas KG, Yanina Pepino M, et al. Nutrition impact symptoms and associated outcomes in post-chemoradiotherapy head and neck cancer survivors: a systematic review. J Cancer Surviv. 2018;12(4):479–94.

Engin A. Obesity-associated breast cancer: analysis of risk factors. Adv Exp Med Biol. 2017;960:571–606.

Hu C. UCLA Center for East-West Medicine. Juicing: body cleansing or nutrient depleting? 2019. exploreim.ucla.edu/wellness/juicingbody-cleansing-or-nutrient-depleting/#-Cleansing2019. Accessed 10 Aug 2020.

Inglis JE, Lin PJ, Kerns SL, Kleckner IR, Kleckner AS, Castillo DA, Mustian KM, Peppone LJ. Nutritional interventions for treating cancer-related fatigue: a qualitative review. Nutr Cancer. 2019;71(1):21–40.

Julia EI, Lin P-J, Sarah LK, Ian RK, Amber SK, Daniel AC, Karen MM, Luke JP. Nutritional interventions for treating cancer-related fatigue: a qualitative review. Nutr Cancer. 2019;71 (1):21–40. https://doi.org/10.1080/01635581.2018.1513046.

Lien YH. Juicing is not all juicy. Am J Med. 2013;126:755–6.

Patient Page. Answers to questions often asked by cancer survivors about nutrition and physical activity. CA Cancer J Clin. 2003;53(5):303–9.

Patient Page. A healthy diet and physical activity help reduce your cancer risk. CA Cancer J Clin. 2012;62:68–9.

Perez-Hernandez AI, Catalan V, GomezAmbrosi J, et al. Mechanisms linking excess adiposity and carcinogenesis promotion. Front Endocrinol (Lausanne). 2014;5:65.

Ravindran R, Jaiswal AK. Wholesomeness and safety aspects of irradiated foods. Food Chem. 2019;285:363–8.

Rinninella E, Mele MC, Cintoni M, et al. The facts about food after cancer diagnosis: a systematic review of prospective cohort studies. Nutrients. 2020;12(8):2345.

Rock CL, Thomson C, Gansler T, et al. American Cancer Society guideline for diet and physical activity for cancer prevention. CA Cancer J Clin. 2020;70(4):245–71.

Tavernise S. Misconception: juice cleansing can remove toxins from your system. New York Times. April 20, 2016;§D:2. 2016. nytimes.com/2016/04/21/health/juice-cleanse-toxinmisconception.html?searchResultPosition=3. Accessed 15 July 2020.

Totmaj AS, Emamat H, Jarrahi F, Zarrati M. The effect of ginger (Zingiber officinale) on chemotherapy induced nausea and vomiting in breast cancer patients: a systematic literature review of randomized controlled trials. Phytother Res. 2019;33(8):1957–65. https://doi.org/10.1002/ptr.6377.

US Food and Drug Administration. Questions & answers on food from genetically engineered plants. 2018. fda.gov/food/food-newplant-varieties/questions-answers-foodgenetically-engineered-plants. Accessed 28 Sept 2020.

World Cancer Research Fund/American Institute for Cancer Research. Diet, nutrition, physical activity and cancer: a global perspective. Continuous Update Project. The Third Expert Report. American Institute for Cancer Research; 2018. wcrf.org/dietandcancer. Accessed 13 Aug 2020.

Yuan C, Babic A, Khalaf N, et al. Diabetes, weight change, and pancreatic cancer risk. JAMA Oncol. 2020;6(10):e202948.

Chapter 4
Dietary Recommendations of Worldwide Cancer-Related Organizations

Abstract The issue of providing dietary guidelines to cancer survivors is important owing to the influence of dietary factors in cancer patients undergoing treatment and also after drug holidays. Several National and International agencies, which are involved in cancer research and public health care, have taken lead in providing diet-related information and guidance for cancer survivors. Worldwide, series of clinical, preclinical and epidemiological reports are accumulating in logarithmic fashion linking the role of diet, dietary patterns, ingredients and nutrition with cancer survival and mortality. Moreover, the notable cancer-related worldwide national and international organizations are providing dietary and nutritional guidelines for the cancer patients. For example, World Cancer Research Fund International (WCRF: wcrf.org) in collaboration with American Institute for Cancer Research (AICR) is running a programme called continuous update project (CUP), wherein the worldwide experts are consistently analysing and interpreting the diet- and cancer-related scientific literature (clinical, epidemiological, cohort, observational and case studies) and providing the evidence-based recommendations in relation to diet, nutrition and physical activity in concert with cancer risk and survival of different types of cancer patients across the world. Although the present global profile dietary recommendation is consistently inconsistent, however, the ultimate aim of dietary research is to provide the patient-centric dietary information/guidelines, which at present is lacking to a greater extent.

Keywords Dietary guidelines · Recommendations · Cancer survivors · Treatment

Abbreviations

ACS	American Cancer Society
AICR	American Institute for Cancer Research
ASN	American Society for Nutrition
ASN	American society for nutrition
CCS	Canadian Cancer Society
CRUK	Cancer Research UK

R. N. Gacche, *Dietary Research and Cancer*,
https://doi.org/10.1007/978-981-16-6050-4_4

CSC	Cancer Support Community
CUP	Continuous Update Project
ECAC	The European Code against Cancer
EPIC	European Prospective Investigation into Cancer and Nutrition
ESPEN	The European Society for Clinical Nutrition and Metabolism
LNCE	Limited no conclusive evidence
LSE	Limited suggestive evidence
NCCN	National Comprehensive Cancer Network
NCI	National Cancer Institute
SCE	Strong convincing evidence
SPE	Strong probable evidence
UICC	Union for International Cancer Control
WCRF	World Cancer Research Fund
WHO	World Health Organization
WHO: ECAC	European Code Against Cancer

Introduction

Overview of Dietary Recommendations of Cancer-Related Agencies

Several National and International agencies, which are involved in cancer research and public health care, have taken lead in providing diet-related information and guidance for cancer survivors (Table 4.1). Worldwide, series of clinical, preclinical and epidemiological reports are accumulating in logarithmic fashion linking the role of diet, dietary patterns and nutrition with cancer survival and mortality. Moreover, the notable cancer-related worldwide national and international organizations are providing dietary and nutritional guidelines for the cancer patients. For example, World Cancer Research Fund International (WCRF: wcrf.org) in collaboration with American Institute for Cancer Research (AICR) is running a programme called continuous update project (CUP), wherein the worldwide experts are consistently analysing and interpreting the diet- and cancer-related scientific literature (clinical, epidemiological, cohort, observational and case studies) and providing the evidence-based recommendations in relation to diet, nutrition and physical activity in concert with cancer risk and survival of different types of cancer patients across the world. Healthy weight, physical activity, diet rich in whole grains, vegetables, fruits, beans and limiting the inclusion of red meat, processed meat, fast foods, processed foods, foods high in fats, starches or sugars, limiting consumption of sugar drinks, alcohol and avoiding the supplements for cancer prevention are some of the key recommendations recently advocated by WCRF-AICR CUP Report 2018 (WCRF: Continuous Update Project Expert Report 2018). Apart from the general dietary recommendations, the most welcome part of the WCRF-AICR CUP Report 2018 is the

Table 4.1 Comparative profile of general dietary recommendations of national and international agencies for cancer patients

Name of the organization	Recommended food types for cancer patients							Online[a] access/references
	Vegetables, fruits, food grains and protein food	Nonvegetarian diet and fatty foods	Dairy products	Food supplements, micronutrients and vitamins	Drinks/liquid foods	Alcohol	Foods/ ingredients that may increase the risk of cancer	
WCRF-AICR	• Fill at least two-thirds of daily plate (at least 400 g) with whole grains, vegetables, fruits, leguminous pulses, beans, lentils, etc. • Consume 30 g of dietary fibre/ day • Eat nonstarchy vegetables, fruits and pulses instead of starchy roots or tuberous • Limit consumption of high starch, sugar and salty or salt-preserved foods	• Eat moderate amount (350–500 g/ week) of red meat (beef, pork, lamb) and little processed meat • Limit consumption of processed foods high in fat, fast foods and prepared dishes, snacks, etc.	• Avoid consumption of high-fat processed bakery products like white flour bread, pizza, cake, pastries, cookies and candy	• Do not use high-dose dietary supplements for cancer prevention • Avoid micronutrient supplement as it can be achieved through diet	• Do not consume sugar-sweetened drinks •Drink mostly water and unsweetened drinks	• Limit one drink (with 10–15 g of ethanol per serving) for women and two for men/day	• Do not consume Cantonese-style salted fish or grilled fish • Excess red meat or grilled meat, aided salt • Foods containing haem iron	• www.wcrf.org/sites/default/files/Recommendations • Continues Update Project Expert Report 2018. Available at: dietandcancerreport.org "HEAL Well: A Cancer Nutrition Guide" by AICR. Available at www.aicr.org/assets/docs/pdf/education/heal-well-guide.pdf
NCI	• Lots of fruits and vegetables, including raw and cooked vegetables, fruits and fruit juices	• Eat small portions (about 6–7 ounces each day) of lean meat and poultry without	• Low-fat milk products, nonfat yoghurt and skim or 1% milk • Small	• Vitamin C, probiotics, melatonin, oral glutamine	• Drink milkshakes, smoothies, juices or soups in case you cannot eat	• Limit: One drink per day for women and up to two drinks per day for men	• Excess alcohol, red meat, charred meat, dietary acrylamide, high antioxidants,	• www.cancer.gov/publications/patient-education/eatinghints; • www.cancer.gov/about-cancer/treatment/side-effects/appetite-loss/

(continued)

Table 4.1 (continued)

Name of the organization	Recommended food types for cancer patients							Online[a] access/references
	Vegetables, fruits, food grains and protein food	Nonvegetarian diet and fatty foods	Dairy products	Food supplements, micronutrients and vitamins	Drinks/liquid foods	Alcohol	Foods/ ingredients that may increase the risk of cancer	
	• Oats, brown rice or other whole grains and cereals • Beans, peas, nuts and lentils • Eat plenty of protein-rich high-calorie diet	skin • Limit red meat to 18 ounces per week • Small amounts of mayonnaise, fried foods and other high-calorie fat foods	amounts of butter, desserts, etc.				high calcium, artificial sweeteners, excess salt, etc.	nutrition-pdq • www.cancer.gov/about-cancer/causes-prevention/risk/diet
ACS	• Eat at least 2½ cups of variety of colourful vegetables and fruits each day and include at every meal and snacks • Choose 100% vegetable or fruit juices. Eat beans and peas • Avoid sauces, dressings and dips with fruits and vegetables • Choose	• Limit intake of processed meats such as bacon, sausage, lunch meats and hot dogs • Choose fish, poultry instead of red meat • If you eat red meat, choose lean cuts and eat smaller portions	• Prefer low-fat dairy products • Eat protein-rich yoghurt, milk and cheese • Avoid calorie-dense dairy products such as ice cream, donuts, pastries and candy	• Eat dietary antioxidants and not as supplements • Eat recommended levels of calcium and vitamin D • Taking of antioxidant supplements or vitamin-enhanced foods is not recommended	• Drink lower calorie beverages • Limit high-sugar-sweetened beverages such as soft drinks, sports drinks and fruit-flavoured drinks	• "It is best not to drink alcohol" • Two drinks per day for men and 1 drink per day for women	• Excess alcohol • Red and processed meats	• www.cancer.org/treatment/survivorship-during-and-after-treatment • https://www.cancer.org/healthy/eat-healthy-get-active/acs-guidelines-nutrition-physical-activity-cancer-prevention/guidelines.html

	whole-grain breads, pasta and cereals (such as barley, oats and brown rice)	• Avoid meat, poultry and fish by frying or charbroiling		while chemo- or radiation therapy				
WHO: ECAC	• Eat plenty of whole grains, pulses, vegetables, fruits and foods rich in fibres • Include pulses like beans, peas or lentils; like rice, pasta or bread (whole grain) and nuts and seeds • Limit high-calorie foods (foods high in sugar and avoid sugary drinks) • Avoid fast foods and foods high in salt • Not to exceed 5–6 g of salt per day	• Limit high-calorie foods (foods high fat) • Avoid processed meat; limit red meat (<500 g/week)	N/A	• Dietary supplements are not recommended for cancer prevention • It is best to have a diet of ordinary foods that will provide all the nutrients	• Avoid sugary drinks	• Overall, the consumption of <20 and <10 g of pure alcohol/day by men and women, respectively, reduces the cancer risk by 6% than people with higher alcohol intakes • No alcohol is better for cancer prevention	• Excess alcohol • Processed meats • High salt	• http://cancer-code-europe.iarc.fr/index.php/en/ecac-12-ways/diet-recommendation
NCCN	• Eat a minimum of five servings (one-half plate) of fruits and vegetables like dark leafy greens or berries in a day	• Choose lean protein (one-fourth plate) from fish, poultry and tofu • Limit red	• Keep dairy low fat. Select skim milk, low-fat yoghurt and reduced fat cheeses • Avoid	• Food—not supplements—is the best source of vitamins and minerals • Eat foods	• Reduce intake of sweets, and sugary juices, drinking soda that has sugar	• Men: no more than two drinks and women: no more than one drink	• Red meat promotes inflammation in human tissue, and this may stimulate the growth of	www.nccn.org/patients/resources/life_after_cancer/nutrition.aspx

(continued)

Table 4.1 (continued)

Name of the organization	Recommended food types for cancer patients							Online[a] access/references
	Vegetables, fruits, food grains and protein food	Nonvegetarian diet and fatty foods	Dairy products	Food supplements, micronutrients and vitamins	Drinks/liquid foods	Alcohol	Foods/ ingredients that may increase the risk of cancer	
	• Use plant-based seasonings like parsley and turmeric • Go for whole grains (one-fourth plate) and high-fibre breads and cereals, including brown rice, barley, bulgur and oats	meat and processed meats	refined foods, such as donuts and white bread, and those high in sugar	high in vitamin D (salmon, sardine, fortified orange juice, milk, etc.)			cancerous tumours • High-dose supplements may increase cancer risk	
ESPEN	• Diet based on vegetables, fruits and whole grains	• Diet low in saturated fat, red meat	N/A	• Oral nutritional supplements like multivitamins, fish oil supplements (advanced cancer) are recommended to avoid nutritional deterioration	N/A	• Limit alcohol intake	• High intake of red meat (beef, pork, mutton)	• www.espen.org/guidelines-home/espen-guidelines
CRUK	• Eat more fruit and vegetables • Healthy proteins from pulses	• One fourth of plate should be fresh chicken, fish or	• Milk and dairy are good sources of calcium and	• Supplements do not substitute for a healthy diet, so	• Limit sugary drinks as it helps to avoid extra weight	• Both men and women are advised to drink no more than	• Processed and red meat • Alcohol • High sugar	• www.cruk.org/diet

	(e.g. lentils or beans) • Of the one plate, one half should be salad or vegetable, beans, etc., and one-fourth of potato, brown rice, bread or pasta • Eat foods high in fibre (30 g/day), particularly whole grains or whole-grain versions of bread, pasta or rice	eggs • Limit processed and red meat 70 g or less/day • Limit high-calorie foods.	protein, which are needed as part of a healthy, balanced diet • There is evidence that dairy products could reduce the risk of bowel cancer, but there is no strong evidence to any other types of cancer	eat a healthy and balanced diet • Everyone should consider taking vitamin D	gain and keeping a healthy weight cuts the risk of 13 types of cancer	14 units of alcohol a week (1 unit equals 8 g of alcohol)		
CSC	• Eat a diet rich in whole grains, vegetables, fruits (colourful) and beans • Limit processed foods high in starches and sugars • Eat more protein from sources like meat, poultry, fish, eggs, dairy, beans and nuts	• Limit red meats (beef, pork and lamb). Eat not more than 12–18 ounces per week • Limit processed meats (ham, sausage, bacon, salami and hot dogs) • Limit "fast foods" and other processed foods high in fat	• Low-fat dairy products, like yoghurt	• Do not use supplements for cancer prevention. A healthy diet is the best way to get nutrition	• Limit consumption of sugar-sweetened drinks • Drink mostly water and unsweetened drinks	• It is best not to drink alcohol • 1 drink per day for a woman and 2 drinks per day for a man • Alcohol is linked to six different cancers	• Red and processed meats and alcohol	• www.cancersupportcommunity.org/diet-nutrition

(continued)

Table 4.1 (continued)

Name of the organization	Recommended food types for cancer patients							Online[a] access/references
	Vegetables, fruits, food grains and protein food	Nonvegetarian diet and fatty foods	Dairy products	Food supplements, micronutrients and vitamins	Drinks/liquid foods	Alcohol	Foods/ ingredients that may increase the risk of cancer	
CCS	• Eat more snacks of fruit (fresh, frozen, canned or dried) and vegetables (raw, cooked, juices) • Include whole grains, nuts, wheat germ or granola, dried beans, peas, lentils and soya foods	• Eat proteins from fish, poultry, lean red meat, eggs, etc.	• Include dairy products like cheese (hard or cream), butters, cakes or cookies made with whole grains and yoghurt • Dietary supplement may improve drug effectiveness	• Commercial nutrition supplements of vitamins, minerals and others are recommended which is right for patients	• N/A	• Choose to have a alcohol drink after talking advice from healthcare team • Limit alcohol as it may reduce your appetite	• Alcohol also increases the risk of developing certain types of cancer	• www.cancer.ca/en/cancer-information/cancer-journey/living-with-cancer/nutrition-for-people-with-cancer
WHO	• Eat fruit, vegetables, legumes (e.g. lentils and beans), nuts and whole grains (e.g. unprocessed maize, millet, oats, wheat and brown rice) • At least 400 g (i.e. five portions) of fruit and vegetables per day (excluding	• Less than 30% of total energy intake from fats • Unsaturated fats are preferable to saturated and all kinds of trans-fats • It is recommended that the intake of saturated	• Eat reduced fat dairy foods • Limit the consumption of baked foods, cakes, pies, cookies and biscuits containing industrial trans-fats	• The WHO has established Vitamin and Mineral Nutrition Information System (VMNIS) to strengthen the global surveillance of vitamin and mineral deficiencies	• Limit sugary drinks like sweetened beverages, soft drinks, fruit or vegetable juices, powder concentrates, flavoured water, sports drinks, ready-to-drink tea, coffee and	• WHO describes alcohol as toxic and psychoactive substance, which contributes to three million global deaths, disabilities, premature mortality and poor health	• WHO has classified processed meats including ham, bacon, salami and frankfurts as group 1 carcinogens	• www.who.int/en/news-room/fact-sheets/detail/healthy-diet Healthy diet-2019 (https://apps.who.int/iris/handle/10665/325828) • www.who.int/features/qa/cancer-red-meat/en/

	potatoes, sweet potatoes, cassava and other starchy roots) • Less than 10% of total energy intake from free sugars • Less than 5 g of iodized salt/day	fats (<10%) and trans-fats (<1%) of total energy intake			flavoured milk drinks			

WHO World Health Organization, *WCRF-AICR* World Cancer Research Fund–American Institute for Cancer Research, *NCI* National Cancer Institute, *ASN* American Society for Nutrition, *ACS* American Cancer Society, *CSC* Cancer Support Community, *CCS* Canadian Cancer Society, *WHO: ECAC* European Code Against Cancer, *NCCN* National Comprehensive Cancer Network, *ESPEN* The European Society for Clinical Nutrition and Metabolism, *CRUK* Cancer Research UK, *N/A* not available

[a]Accessed in September 2020

comprehensive chapter-wise dietary recommendations in relation to 18 different cancers (Table 4.2) and a comprehensive chapter-wise discourse in relation to six different diet-related risk factors with variety of cancers (https://www.wcrf.org/dietandcancer/recommendations-about). Of note, every section and chapter of WCRF-AICR 2018 diet recommendation in relation to different cancer types and risk factors are described with comprehensive literature citation, which perhaps makes it more meaningful and value addition towards implementing in clinical practice. "HEAL Well: A Cancer Nutrition Guide" (www.aicr.org/assets/docs/pdf/education/heal-well-guide) is a joint publication of AICR, the LIVESTRONG Foundation and Savor Health. This guide is a ready to use reference for diet, nutrition and cancer management. It answers common questions about diet, nutrition and physical activity during and after cancer treatment, and also offers suggestions for amelioration of treatment-related side effects.

According to WHO (www.who.int), approximately one-third of cancer deaths are associated with dietary risks and the good news is that about 30–50% of all cancer cases could be managed and the dietary modifications are an important approach to cancer management (WHO Cancer Prevention). The data further summarize the low levels of dietary inclusions of fruit and vegetable as one of the culprits for the development of cancer (WHO: Cancer: Key Facts, 2019). Moreover, although not exclusively related to cancer prevention, a recent (2019) booklet of WHO on "Healthy Diet" (online: apps.who.int/iris/handle/10665/325828) is a guideline for healthy diet, which perhaps may help cancer survivors in protecting against malnutrition in all its forms (Healthy Diet. Cairo: WHO Regional Office for the Eastern Mediterranean; 2019). Cancer Support Community (CSC) is the largest nonprofit global network of over 175 locations, professionally provides cancer-related knowledge including diet and nutrition and has categorically provided information for the management of side effects of cancer treatments through dietary interventions (www.cancersupportcommunity.org). The booklets titled "Frankly Speaking about Cancer: Eating Well for Cancer Survivors" (www.CancerSupportCommunity.org/SurvivorNutrition) and "Frankly Speaking about Cancer: Eating Well with Head and Neck Cancer" (www.CancerSupportCommunity.org/headneck) by CSC are informative and easy to understand guidance resources for managing diet, nutrition and preparing different food recipes for cancer survivors and head and neck cancer patients, respectively.

European Prospective Investigation into Cancer and Nutrition (EPIC) is one of the largest cancer-related cohort studies in the world dedicated for investigating the association of diet, nutritional status, lifestyle and environmental factors with incidence of cancer and other chronic diseases. EPIC recruits over half a million volunteers from ten European countries study follow-up to almost 15 years (EPIC: http://epic.iarc.fr/). Series of EPIC cohort studies have been published recently; for example, in a cohort of over half a million participants, it was observed that intake of fruit juice was positively associated with risk of differentiated thyroid carcinoma (possibly because of sugar), while no significant association was observed with intake of fruit and vegetables (Zamora-Ros et al. 2018). In another recent EPIC cohort study of 142,196 men, it was observed that consumption of total, caffeinated

Table 4.2 WCRF-AICR dietary guidelines for different types of cancer (continues Update Project Report 2018)

Type of cancer	Dietary inclusions that decrease the cancer risk	Dietary inclusions that increase the cancer risk	Online access
Bladder	**LSE**: Vegetables and fruits, tea	**SPE**: Arsenic in drinking water	www.wcrf.org/dietandcancer/bladder-cancer
	LNCE: Cereals, pulses, meat, poultry, fish, fat, milk, yoghurt, cheese, soft drinks, fruit juices, coffee, green tea, caffeine, alcohol, sweeteners, vitamins A, C and E, calcium, folate, selenium, β and α-carotene, lycopene, lutein, zeaxanthin, flavonoids, tocopherol, multivitamin supplements, etc.		
Breast	**LSE:** Nonstarchy vegetables (ER breast cancer), dairy products, foods containing carotenoids and calcium	**SPE**: Alcoholic drinks	www.wcrf.org/dietandcancer/breast-cancer
	LNCE: Cereals, dietary fibres, nonstarchy vegetables (ER^+ breast cancer), fruits, pulses, soya products, red and processed meat, poultry, fish, eggs, fats and oils, mono- and polyunsaturated fatty acids, trans-fatty acids, sugary foods and drinks, coffee tea, vitamins A, B6, B12, C, D and E, riboflavin, calcium supplements, iron, selenium, phytoestrogens, isoflavones, dietary patterns, etc.		
Cervical	**LSE:** Carrots	Not identified	www.wcrf.org/dietandcancer/cervical-cancer
	LNCE: Nonstarchy vegetables, fruits, milk, retinol, vitamin E, alcoholism, body fatness		
Colorectal	**SPE:** Whole grains, foods containing dietary fibres, dairy products, calcium supplements **LSE:** Foods containing vitamin C, fish, vitamin D, multivitamin supplements	**SCE:** Processed meat and alcoholic drinks **LSE**: Low intake of nonstarchy vegetables and fruits, foods containing haem iron	www.wcrf.org/dietandcancer/colorectal-cancer
	LNCE: Cereal grains, potatoes, animal fat, poultry, shellfish seafood, cholesterol, legumes, garlic, nondairy sources of calcium, added sugars, coffee, tea, caffeine, folate, vitamins A, B6 and E, selenium, beta-carotene, lycopene, retinol, dietary pattern		
Endometrial	**SPE:** Coffee	Not identified	www.wcrf.org/dietandcancer/endometrial-cancer
	LNCE: Cereals, fruits, vegetables, soya and soya products, red meat, processed meat, poultry, fish, eggs, milk, dairy products, dietary fibre, total fat, saturated fatty acids, cholesterol, animal fat, tea, beta-carotene, folate, vitamins C and E, multivitamins, alcohol, dietary pattern		

(continued)

Table 4.2 (continued)

Type of cancer	Dietary inclusions that decrease the cancer risk	Dietary inclusions that increase the cancer risk	Online access
Gallbladder	**LNCE:** Peppers (capsicums), fish, coffee, tea, alcohol, sugar, vitamin C, calcium and vitamin D supplements, low-fat diets		www.wcrf.org/dietandcancer/gallbladder-cancer
Kidney	**SPE:** Alcoholic drinks	**LSE:** Arsenic in drinking water	www.wcrf.org/sites/default/files/Kidney-cancer
	LNCE: Cereals and their products, dietary fibres, vegetables, fruits, fish, eggs, milk, and dairy products, total fats, soft drinks, tea, coffee, calcium, vitamins A, C, E and B6, beta- and alpha-carotene, lycopene, beta-cryptoxanthin, lutein, zeaxanthin, Adventist diet		
Liver	**SPE:** Coffee **LSE:** Fish	**SCE:** Alcoholic drinks	www.wcrf.org/dietandcancer/liver-cancer
	LNCE: Cereal grains and their products, nonstarchy vegetables, fruits, peanuts 9ground nuts), meat and poultry, salted fish, tea green tea, calcium and vitamin D supplements, vitamin C, low-fat diet		
Lung	**LSE:** Vegetables, fruits, foods containing carotenoids, beta-carotene, vitamin C, isoflavones	**SCE**: High-dose beta-carotene supplements, arsenic in drinking water **LSE:** Red meat and processed meat, alcoholic drinks	www.wcrf.org/sites/default/files/Lung-cancer
	LNCE: Cereals (grains, starchy tubers, vegetables and fruits) (never smoker), dietary fibre, legumes, citrus fruits, poultry, fish, egg, dairy products, total meat, total fat, animal fat, plant oils, soft drinks, coffee, tea, vitamin A, selenium, thiamine, riboflavin, niacin, vitamin B6, folate, foods containing vitamin C, vitamin E, beta-carotene, calcium, copper, lycopene, lutein, zeaxanthin, isoflavones, multivitamin supplements, patterns of diet, etc.		
Mouth, pharynx and larynx	**LSE**: Nonstarchy vegetables, healthy diet patterns, coffee	**SCE:** Alcoholic drinks **LSE:** Mate (an infusion prepared from dried leaves (*Ilex paraguariensis*)	www.wcrf.org/dietandcancer/mouth-pharynx-larynx-cancers
	LNCE: Cereals (grains), starchy roots, tubers, plantains, fruits, legumes, dietary fibres, total red and processed meat, poultry, fish eggs, dairy products, total and animal fat, plant oils, tea, soft drinks, fruit juices, vitamins A, C, D and E, riboflavin, calcium, selenium, carotenoids, iron, thiamine, folate, etc.		
Nasopharyngeal	**LSE**: Nonstarchy vegetables, fruits	**SPE**: Cantonese-style salted fish	www.wcrf.org/dietandcancer/nasopharyngeal-cancer

(continued)

Table 4.2 (continued)

Type of cancer	Dietary inclusions that decrease the cancer risk	Dietary inclusions that increase the cancer risk	Online access
	LNCE: Cereals, nuts, seeds, herbs, spices, condiments, meat, fish, shellfish, seafoods, plant oils, tea, alcohol, salted plant food, Chinese-style pickled cabbage, pickled radish, pickled mustard leaf, Chinese-style preserved salted eggs, fermented tofu and soya products		
Oesophageal	**LSE:** Fruits and vegetables	**SCE:** Alcoholic drinks; **SPE**: Mate; **LSE**: Processed meat	www.wcrf.org/dietandcancer/oesophageal-cancer
	LNCE: Dietary fibre, fruit, red, processed and total meat, poultry, fish, coffee, mate, alcohol, vitamins C and E, folate, beta-carotene, patterns of diet, cereals, starchy roots, tubers and plantains, pulses, legumes, soya and products, herbs, spices and condiments, milk & dairy products, total fat, saturated and mono- and polyunsaturated fatty acids, sugary foods and drinks, salt, salting, fermented, soaked and cured foods, nitrates and nitrites, frying, grilling, barbecuing, vitamin A, thiamine, riboflavin, calcium, iron, zinc, pro-vitamin A carotenoid, beta-cryptoxanthin, etc.		
Ovarian	**LNCE:** Vegetables, fruits, pulses, legumes, red and processed meat, vegetarian and individual-level dietary pattern, coffee, tea, dietary fibre, total fat, saturated, mono- and polyunsaturated fatty acids; vegetable, animal and trans fats, dietary cholesterol, alcohol, vitamins A and C, serum vitamin D, lycopene, lactose, calcium, etc.		www.wcrf.org/dietandcancer/ovarian-cancer
Pancreatic	Not identified	**LSE:** Red and processed meat, alcoholic drinks (heavier drinking), foods and beverages containing fructose, foods containing saturated fatty acids	www.wcrf.org/dietandcancer/pancreatic-cancer
	LNCE: Fruits, vegetables, folate, fish, eggs, tea, soft drinks, coffee, sucrose, total fat, mono- and polyunsaturated fats, dietary cholesterol, vitamin C, multivitamin/mineral supplements		
Prostate	Not identified	**LSE:** Dairy products, diet high in calcium, low plasma alpha-tocopherol and selenium concentrations	www.wcrf.org/dietandcancer/prostate-cancer
	LNCE: Cereals, dietary fibres, potatoes, nonstarchy vegetables, fruits, pulses, legumes, processed and red meat, poultry, fish, eggs, total fat, saturated fatty acids, mono- and polyunsaturated fatty acids, plant oils, sucrose, sugary foods and drinks, coffee, tea and alcoholic drinks,		

(continued)

Table 4.2 (continued)

Type of cancer	Dietary inclusions that decrease the cancer risk	Dietary inclusions that increase the cancer risk	Online access
	vitamin A, alpha-carotene, lycopene, riboflavin, folate, thiamine, vitamin C, D and E supplements, multivitamins, selenium supplements, iron, calcium supplements, vegetarian diets, dietary patterns, etc.		
Skin	**LSE:** Coffee (basal cell carcinoma, malignant melanoma)	**SPE:** Arsenic in drinking water; **LSE:** Alcohol (basal cell carcinoma, malignant melanoma)	
	LNCE: Potatoes, nonstarchy vegetables, fruits, milk, coffee, decaffeinated coffee, tea, alcoholic drinks, total fat, cholesterol, retinol in diet and supplements, beta-carotene in the diet, vitamins A, C, D and E, selenium, patterns of diet, meat, processed meat, fish, offal, poultry, eggs, all vegetables, multivitamin supplements, lycopene, lutein, zeaxanthin, carotenoids, alpha-carotene, high dose of beta-carotene supplements has strong substantial unlikely effect on risk of skin cancer		www.wcrf.org/dietandcancer/skin-cancer
Stomach	**LSE:** Citrus fruit (cardia)	**SPE:** Alcoholic drinks, food preserved by salting **LSE:** Processed meat (noncardia), grilled or barbecue meat and fish, low fruit intake	www.wcrf.org/dietandcancer/stomach-cancer
	LNCE: Cereal grain, dietary fibre, vegetables, legumes, potatoes, starchy roots, tubers, citrus fruit (noncardia), nuts and seeds, herbs, chilli, spices, condiments, meta- and processed meat, poultry, fish, eggs, milk, dairy products, salt, added salt, fruit juices, sugars, coffee, tea, green tea, fried or dried food, dietary nitrate and nitrite, fats and oils, fatty acid composition, retinol, beta-carotene, cholesterol, vitamins C and D; riboflavin, thiamine, multivitamin/mineral supplements, calcium, iron, selenium		

SCE strong convincing evidence, *SPE* strong probable evidence, *LSE* limited suggestive evidence, *LNCE* limited no conclusive evidence

or decaffeinated coffee or tea was not associated with risk of total prostate cancer or cancer at any stage or grade (Sen et al. 2019). The European Code Against Cancer (ECAC): an initiative of the European Commission and having linkage with WHO has described the 12 ways of reducing the cancer risk, which also includes the dietary interventions (www.cancer-code-europe.iarc.fr). Eating plenty of whole grains, vegetables, fruits, pulses, limiting high-calorie foods, sugary drinks, processed meat, red meat, high salt and alcohol are some of the key features of ECAC dietary guidelines (Schüz et al. 2015).

The European Society for Clinical Nutrition and Metabolism (ESPEN) is a devoted society that studies the metabolic problems associated with acute diseases and their nutritional implications and management. ESPEN has systematically drafted the evidence-based guidelines on nutrition for cancer patients (Arends et al. 2017). Union for International Cancer Control (UICC) with a network of 171 member countries is basically involved in convening, capacity building and advocacy towards reducing the global cancer burden (www.uicc.org). As such, UICC has not formulated any dietary guidelines for cancer management, but UICC official journal like "International Journal of Cancer" has series of epidemiological cohort studies, RCTs, related to dietary interventions and cancer prevention (Rada-Fernandez de Jauregui et al. 2018; He et al. 2019).

American Cancer Society (ACS) has drafted evidence-based diet guidelines for cancer prevention, which strongly recommends the adherence to dietary recommendations for minimizing cancer mortality and increasing cancer patient survival. Consumption of plant-based healthy diet, choosing appropriate amount of foods and beverages for maintaining healthy weight, limiting consumption of processed meat and red meat, consuming minimum of 2.5 cups of vegetables and fruits each day, prioritizing whole grains and limiting alcohol (1 drink/day for women and 2 drinks/day for men) are some of the highlights of ACS dietary guidelines for cancer prevention. The guidelines also address the diet and nutrition issues during and after cancer treatment (Kushi et al. 2012). The recently published ACS guidelines towards the intake of diet and need of physical activity for cancer management clearly reiterated the need of consuming healthy dietary patterns, which includes the dietary sources that are rich in nutrients required to achieve and sustain a healthy body weight. The ACS guidelines 2020 also re-strengthened the inclusion of variety of vegetables, especially dark green, orange and red in colours, eating of beans and peas which are rich in fibres, and intake of coloured whole fruits and grains. The guidelines once again discourage the intake of red and processed meats, beverages (sugar-sweetened), highly processed foods and refined grain commercial products. Of note, the ASC guidelines for the first time has stated that "It is best not to drink alcohol", but have also provided the limits of alcoholic drinks for men and women (Rock et al. 2020).

The National Cancer Institute (NCI) has precisely described the nutrition and diet in cancer care emphasizing on the effects of cancer treatment on nutrition, nutrition assessment in cancer care, types of nutrition support, nutrition trends in cancer, diet after and before treatment, etc. (www.cancer.gov). The NCI booklets "Chemotherapy and You: Support for People with Cancer" (www.cancer.gov/publications/patient-education/chemo-and-you) are a patient-friendly information resource that describes the diet-related information in relation to chemotherapy. Another NCI book "Eating Hints: Before, during, and after Cancer Treatment" (www.cancer.gov/publications/patient-education/eating-hints) is a compilation of hints about common types of eating problems of cancer patients and the possible solutions to manage them along with variety of recipes for cancer survivors. American Society for Nutrition (ASN: nutrition.org) has highlighted the significance of diet and nutrition in the management of cancer prevention and treatment. ASN has also described several food categories that are closely associated with increased or decreased risk of cancer. National Comprehensive Cancer Network (NCCN: www.nccn.org) comprising of 21 world's leading cancer centres have specifically given recommendations towards general and numerical dietary macronutrient requirements and food types for cancer survivors before and after cancer treatment. Comparative dietary guideline profiles of major institutes of NCCN show that the guidelines suggested are consistently inconsistent (Champ et al. 2013).

Canadian Cancer Society (CCS: www.Cancer.ca) has outlined the dietary and nutritional recommendations for the people with cancer and also nutritional guidelines for children with cancer. Moreover, the society has categorically mentioned the cancer-specific nutritional requirements and answers to diet-related questions from cancer survivors, which helps the cancer patients to manage their dietary plans before and after cancer treatment (Canadian Cancer Society: Nutrition for People with Cancer, 2019). Among the online availability of various informative booklets on different types of cancers, the resource book "Advanced Cancer" (www.cancer.ca/~/media/cancer.ca/CW/publications/Advanced%20cancer/32043-1-NO.pdf) is an excellent repertoire of suggestions for people with advanced cancer and their caretakers. There are practical and experience-based dietary and other tips for amelioration of variety of symptoms of advanced cancer (Advanced Cancer: Canadian Cancer Society 2015). Cancer Research UK (CRUK: www.cancerresearchuk.org): a cancer research and awareness charity institution in the United Kingdom has provided convincing information about healthy diet and nutrition for cancer patients and also answered the frequently asked diet-related questions by the cancer survivors. For example, if patients face difficulty in swallowing food (side effect of few chemodrugs), CRUK has recommended a soft diet and provided tips and furnished variety of soft diet meals and recipes.

While comparing the dietary guidelines and recommendations provided by several national and international cancer agencies and organizations (Table 4.1), it shows that they are varied and often inconsistent lacking the concrete foundation of rigorous clinical evidence. Besides the small quantitative and qualitative differences in food types, majority of the organizations have consensus over inclusion of fruits, vegetables, whole grains, fibre foods, beans, lentils, plant-based protein-rich diet and limiting the inclusion of red meat, processed meat, high-fat foods, salty foods, sugary drinks and excess alcohol. With variations in recommendations towards dairy products, all the agencies have described red, processed meat and excess alcohol as cancer risk factors. The discrepancies also lie with suggestions for intake of food supplements for cancer prevention. For example, the agencies such as CCS and ESPEN are in favour of inclusion of food supplements so as to avoid the nutritional deterioration. However, the premier organizations like WCRF-AICR, ACS (except calcium and vitamin D) ECAC, NCCN, CRUK and CSC discourage the inclusion of food supplements for cancer prevention and advocate the inclusion of dietary sources of food supplements (Table 4.1). Therefore, there is need to formulate improved, patient-centred, cancer-specific dietary recommendations/guidelines, which perhaps will be value addition towards effective management of treatment modalities leading to improvement in quality of life and patient survival.

Of note, the impact of guidelines provided by the various national and international agencies has been limited owing to moderate interest of clinical oncologists in diet and nutritional interventions (Spiro et al. 2006). Perhaps the guideline confusions and the complex pathophysiology of individual cancer patient might be a factor for moderate interest of oncologists and physicians towards dietary interventions.

Conclusion

The cancer-related organizations/research institutes have seriously undertaken the issue of addressing dietary guidelines to cancer patients; however, the lack of empirical data limits their clinical applications. Owing to differences in dietary patterns, demographic/environmental factors and ethnicity, it is practically difficult to formulate the globally accepted dietary guidelines for cancer patients, but based on the empirical evidence, some of the common "do's and don'ts" can be certainly formulated for variety of cancers. Several national and international agencies, which are involved in cancer research and public health care, have taken lead in providing diet-related information and guidance for cancer survivors. Although the present global profile dietary recommendation is consistently inconsistent, however, the ultimate aim of dietary research is to provide the patient-centric dietary information/guidelines, which at present is lacking to a greater extent. *PS*: The cancer patients should consider the guidelines authentic globally acknowledged cancer-related organizations/research institute and should not be victimized to the growing "evidence-poor" advertisement of diet-mediated cancer prevention.

References

Arends J, Bachmann P, Baracos V, Barthelemy N, Bertz H, Bozzetti F, et al. ESPEN guidelines on nutrition in cancer patients. Clin Nutr. 2017;36(1):11–48.

Champ CE, Mishra MV, Showalter TN, Ohri N, Dicker AP, Simone NL. Dietary recommendations during and after cancer treatment: consistently inconsistent? Nutr Cancer. 2013;65(3):430–9.

He X, Wu K, Zhang X, Nishihara R, Cao Y, Fuchs CS. Dietary intake of fiber, whole grains and risk of colorectal cancer: an updated analysis according to food sources, tumor location and molecular subtypes in two large US cohorts. Int J Cancer. 2019;145(11):3040–51. https://doi.org/10.1002/ijc.32382.

Kushi LH, Doyle C, McCullough M, et al. American Cancer Society guidelines on nutrition and physical activity for cancer prevention: reducing the risk of cancer with healthy food choices and physical activity. CA Cancer J Clin. 2012;62(1):30–67.

Rada-Fernandez de Jauregui D, Evans CEL, Jones P, Greenwood DC, Hancock N, Cade JE. Common dietary patterns and risk of cancers of the colon and rectum: analysis from the United Kingdom Women's Cohort Study (UKWCS). Int J Cancer. 2018;143(4):773–81.

Rock CL, Thomson C, Gansler T, et al. American Cancer Society guideline for diet and physical activity for cancer prevention. CA Cancer J Clin. 2020;70(4):245–71.

Schüz J, Espina C, Villain P, Herrero R, Leon ME, Minozzi S, et al. European Code against Cancer 4th Edition: 12 ways to reduce your cancer risk. Cancer Epidemiol. 2015;39S:S1–S10.

Sen A, Papadimitriou N, Lagiou P, Perez-Cornago A, Travis RC, Key TJ, et al. Coffee and tea consumption and risk of prostate cancer in the European Prospective Investigation into Cancer and Nutrition. Int J Cancer. 2019;144(2):240–50.

Spiro A, Baldwin C, Patterson A, Thomas J, Andreyev HJ. The views and practice of oncologists towards nutritional support in patients receiving chemotherapy. Br J Cancer. 2006;95:431e4.

WCRF (World Cancer Research Fund)/American Institute for Cancer Research. Continues Update Project Expert Report. Recommendations and public health and policy implications. 2018. dietandcancerreport.org. Accessed 24 Sept 2020.

Zamora-Ros R, Béraud V, Franceschi S, et al. Consumption of fruits, vegetables and fruit juices and differentiated thyroid carcinoma risk in the European Prospective Investigation into Cancer and Nutrition (EPIC) study. Int J Cancer. 2018;142(3):449–59.

Chapter 5
Outcome of "Diet and Cancer"-Related Clinical Trials

Abstract The plethora of preclinical research data inspired the scientific community and pharma sector to design and conduct the clinical trials for assessing the dietary interventions in cancer prevention, improvising the outcome of chemotherapy and reducing toxicity and many other aspects related to betterment of cancer patient's life. Over 1000 clinical trials of diet, nutrition, dietary supplements and foods in relation to cancer are registered at clinicaltrials.gov. Unfortunately, none of the clinical effects of dietary interventions have been approved by FDA for cancer prevention. Therefore, the upcoming clinical research settings should be more focused on understanding the molecular mechanisms and mode of action of dietary interventions on tumour/cancer cell metabolism and its impact on current chemotherapy efficacy. Such endeavours may help to develop dietary modifications as adjuvant therapeutic approach for effective management of cancer.

Keywords Diet · Cancer · Clinical trials · Outcome · Cancer patients

Abbreviations

AHR	Aryl hydrocarbon receptor
ARNT	Aryl hydrocarbon receptor nuclear translocator
CASP3	Caspase 3
CD44	Clusters of differentiation 44
CDKN1A	Cyclin-dependent kinase inhibitor 1A
COX-2	Cycloxygenase-2
CTNNB1	Gene encoding for β-catenin protein 1
DCIS	Ductal carcinoma in situ
H3K9 or 18	Histone 9 or 18
HDAC 3 or 6	Histone deacetylase 3 or 6
HF	High fat
HIF1A	Hypoxia-inducible factor 1 subunit alpha
HPGD	Prostaglandin dehydrogenase
IDC	Invasive ductal carcinoma

R. N. Gacche, *Dietary Research and Cancer*,
https://doi.org/10.1007/978-981-16-6050-4_5

JUN	Jun proto-oncogene, a transcription factor subunit
KRAS	(Kirsten rat sarcoma virus) gene
MYC	MYC proto-oncogene, a transcription factor
NFAM1	NFAT-activating protein with ITAM motif 1
NFAT	Nuclear factor of activated T (cells), a transcription factor
NF-kB	Nuclear factor κ B
OSM	Oncostatin M, a pleiotropic cytokine
p21	Cyclin-dependent kinase inhibitor 1
PSA	Prostate-specific antigen
PTGS1	Prostaglandin H synthases
QoL	Quality of life
RCT	Randomized controlled trials
V-rel	A Oncogene product from the avian reticuloendotheliosis virus strain T

Introduction

The emerging preclinical studies accumulated in the recent years have demonstrated that dietary modifications and interventions can help in improving the therapeutic index of current chemotherapy drugs, immunosurveillance in concert with immunotherapy, chemotherapy and radiotherapy and reducing side effects/off-target toxicities of current drugs. These interesting molecular findings have inspired the scientific community and the pharma sector to design and conduct clinical trials for investigating the efficacy of dietary interventions to complement with the current treatment modalities for cancer management (Lévesque et al. 2019).

In the Western countries, the overnutrition has resolved the problem of undernutrition as a matter of societal and medical problem (NCD Risk Factor Collaboration, NCD-RisC 2016). The ultra-processed high-calorie foods (rich in sugars, carbohydrates, fats and salt) and soft drinks have adversely affected the majority of population in the developed countries (Dehghan et al. 2017). On contrary, the "normal dietary pattern" cannot be appreciated as a statistical standard, but should be defined by public guidelines. These guidelines are mainly overshadowed by political influence and arguments of epidemiological, and are different across many countries (World Health Organization WHO Global nutrition policy review 2016–2017 2018). Despite all arguments and limitations, it is well accepted that the quality and quantity of diet and nutrition play a significant role in the development of cancer risk (Key et al. 2002).

Conducting randomized controlled clinical trials (RCTs) signifies an authentic evidence for the integration of dietary agents in clinical practice. Of note, nonrandomized studies frequently over predict the magnitude of the dietary impact by a margin of 50–100% (Meyskens and Szabo 2005). Establishing clinical evidence towards dietary interventions in the prevention of cancer has remained one of the thrust areas in the mainstream research (Lévesque et al. 2019). Till date, over 1000 clinical trials for diet, nutrition, dietary supplements and foods are available at clinicaltrials.gov. More precisely, when a search was made using query words like "diet & cancer", "nutrition & cancer", "food & cancer" and "supplement & cancer", until end of September 2020, there were over 1069 clinical trials registered at Clinicaltrials.gov. Of the registered trials, over 509 were completed, 70 trials have been terminated, 29 studies have been withdrawn and over 228 trials are being recruited. The statistics clearly focus on the evolving interest of validating the impact of dietary interventions in cancer prevention.

Outcome of "Diet and Cancer"-Related Clinical Trials

In fact, comparatively a small number of clinical studies have been published in the peer-reviewed literature resources, which mainly focus on the efficacy of dietary interventions and improvement of the cancer therapy. Though there are hundreds of clinical trials with completed status on ClinicalTrials.gov, many of them lack appropriate published information on the outcome of the trials. Series of clinical trials (Clinical trials.gov) have been conducted for validating the adjuvant or combinatorial therapeutic significance of well-studied dietary metabolites like curcumin (NCT00852332: breast cancer, NCT01333917: colorectal cancer, NCT01490996: colonic cancer, NCT02095717: prostate cancer, NCT02064673: prostate cancer, NCT02439385: colorectal cancer, NCT02944578: precancer cervical lesions); resveratrol (NCT00098969: solid tumour, NCT00256334: colon cancer, NCT00433576: adenocarcinoma of the colon); green tea extract (NCT00676793: breast cancer, NCT00303823: cervical cancer, NCT00233935: Barrett oesophagus cancer, NCT02029352: carcinoma); epigallocatechin-3-gallate (NCT02577393: lung neoplasms, NCT02891538: colon cancer); genistein (NCT00118040: bladder cancer; NCT00290758: breast cancer, NCT01036321: prostate cancer, NCT01628471: non-small cell lung cancer); lycopene (NCT01105338, NCT00068731, NCT00006078, NCT00093561, NCT01882985: prostate cancer); sulforaphane (NCT01228084: adenocarcinoma of the prostrate, NCT00894712, NCT00982319, NCT00843167: breast cancer, NCT01950143, NCT01265953: prostate cancer). Apart from the registered studies, there are series of RCTs, and prospective and retrospective observational and cohort studies reported elsewhere (Yang et al. 2016; Lévesque et al. 2019; Cohen et al. 2019; Dinu et al. 2018; de van der Schueren et al. 2018). In fact, establishing a clinical evidence for the dietary interventions in cancer prevention is a challenging task where timing and brevity of intervention, nonlinear diet–cancer association, issues of baseline nutritional status, patient recruitment and magnitudes of the diet are the key factors for outcomes (Mayne et al. 2012, 2016). In the diet–cancer-related clinical trials, there is a gap between the cancer epidemiology and RCTs.

Besides many biases and confounders in designing the clinical trial settings in the mainstream of diet–cancer studies, there is evolving trend of assessing the impact of nutritional and dietary factors in cancer prevention. However, demonstrating the mechanistic relationships between diet and cancer prevention in concert with epidemiological investigations is still a challenging task. A cursory look at the recently completed (Table 5.1) and currently recruited clinical trials (clinicaltrials.gov) shows that many epidemiological and pathophysiological mechanistic advanced parameters have been considered while designing the diet–cancer clinical trials. For example, a cohort study of 44,825 prostate cancer (PC) cases and 27,904 controls with 6-month lycopene and green tea dietary advice or supplementation intervention revealed that consuming dietary lycopene may reduce prostate cancer risk by reducing the levels of circulating pyruvate and the outcome also suggests that the pyruvate may boost PC risk (Beynon et al. 2019). In a systematic prospective cohort setting involving 56,048 human volunteers of the Danish diet and the study duration of 23 years, it was observed that a moderate habitual inclusion of dietary flavonoids was found to be inversely linked with overall cause, cardiovascular and cancer-related mortality. Precisely, the intake of 1000 mg/day of total flavonoid reduced the cancer-related mortality and the consumption of flavonoid-rich functional foods is recommended (Bondonno et al. 2019). The human clinical trial aimed at evaluating the effect of ketogenic diet in advanced cancer across different tumour types concluded that modified Atkins diets were found to be safe in case of advanced cancer patients (Tan-Shalaby et al. 2016). A RCT is designed to find the relationship between broccoli intake with biomarkers for histone acetylation and cell cycle regulation in breast cancer tissue. An inverse relation was noted between the intake of total cruciferous vegetable and the expression Ki-67 protein of breast ductal carcinoma tissue under in situ (Zhang et al. 2016). The series of clinical trials on diet and cancer exhibited the positive correlation of dietary intervention with cancer prevention. For example, low-fat diet effectively increases the efficacy of abiraterone acetate in patients with castration-resistant prostate cancer (Szmulewitz et al. 2018). The unrestricted ketogenic diet was found safe and tolerable by patients having recurrent glioblastoma (Rieger et al. 2014). Moreover, the dietary intake of pomegranate extract (ellagitannin) and fish (long-chain n-3 polyunsaturated fatty acids) influences the regulation of colorectal cancer-related genes and helps reduce the risk of colorectal cancer (Nuñez-Sánchez et al. 2017; Aglago et al. 2019).

Although there is sizable positive correlation of dietary factors in cancer amelioration (Singh et al. 2017), however, the outcome of various trials carried out under different settings clearly shows that there is lack of strong mechanistic dietary evidence for cancer prevention (Table 5.1). In fact, there is an impressive track record of preclinical research demonstrating the promising effects of dietary interventions in cancer prevention; perhaps, the similar kinds of results are lacked in various human clinical trials (Yang et al. 2016; Nosrati et al. 2017). There are series of controversial outcomes of the clinical trials, and some of them are the inclusion of fruits and vegetables that have long been associated with reduced risk of lung and other cancers. However, under clinical settings when higher doses of alpha-tocopherol (50 mg/day) and beta-carotene (20 mg/day) alone or in combinations were administrated to 29,000 Finnish male smokers, a higher frequency of lung cancer was seen in men who have received the beta-carotene than that of control

Table 5.1 Overview of representative completed clinical trials addressing the role of diet, nutrition and food supplements in different types of cancer patients

Author and trial ID	Purpose of clinical trial	Design and description of the trial	Outcome of the trial	Conclusion
Beynon et al. (2019) Not registered	To study the effects of lycopene and green tea on the metabolome of men at risk of prostate cancer	• A randomized controlled trial of 44,825 prostate cancer (PC) cases and 27,904 controls with 6-month lycopene and green tea dietary advice or supplementation intervention on quantitation of 159 serum metabolites	• The testing of 159 serum metabolites level by NMR revealed that increasing dietary lycopene may result in changes in circulating levels of valine, acetate, pyruvate, diacylglycerol and docosahexaenoic acid	• Lycopene supplementation reduced levels of circulating pyruvate, and Mendelian randomization analysis suggests that pyruvate may boost PC risk • Consuming dietary lycopene may reduce prostate cancer risk
Manson et al. (2019) NCT01169259	To investigate the effect of vitamin D supplements in cancer prevention	• A randomized, placebo-controlled trial enrolling 25,871 participants provided with vitamin D3 at a dose of 2000 IU per day, and the end point was incidence of invasive cancer of any type	• Supplementation with vitamin D was not associated with a lower risk of all types of cancer • During a median follow-up of 5.3 years, cancer was diagnosed in 1617 participants (793 in the vitamin D group and 824 in the placebo group)	• Supplementation with vitamin D did not result in a lower incidence of invasive cancer
Szmulewitz et al. (2018) NCT01543776	To assess the efficacy of low dose of abiraterone acetate (AA) with a low-fat meal in patients with castration-resistant prostate cancer (CRPC)	• 72 patients progressive CRPC assigned to standard or low dose of AA • Prostate-specific antigen (PSA) was assessed monthly, and testosterone/dehydroepiandrosterone sulphate was assessed every 12 weeks	• There was a greater effect on PSA in the low-dose arm • The PSA response rate was 58% in low arm and 50% in standard AA arm • Androgen levels decreased similarly in both arms	• Low-dose AA (with low-fat breakfast) is non-inferior to standard dosing with respect to PSA metrics

(continued)

Table 5.1 (continued)

Author and trial ID	Purpose of clinical trial	Design and description of the trial	Outcome of the trial	Conclusion
Nuñez-Sánchez et al. (2017) NCT01916239	To investigate the ellagitannin (ET) containing pomegranate extract (PE) on expression of colorectal cancer (CRC)-related genes in normal and cancerous colon tissues	• A randomized trial of 35 CRC patients who were supplemented with 900 mg of an ET-containing PE/day The patients were evaluated for expression of various CRC-related genes in normal and cancerous colon tissues before (biopsies) and after (surgical specimens) 5–35 days of supplementation	• The gene expression profile indicated the upregulation of MYC, CD44, CDKN1A and CTNNB1, and downregulation of CASP3 gene, while expression of KRAS was not affected • Intake of PE modulated the impact of the protocol on gene expression in a gene- and tissue-specific manner	• Oral intake of an ET-containing PE influences the regulation of CRC-related genes
Marx et al. (2017) ACTRN12613000120774	To determine the effect of time and dose of ginger on chemotherapy-induced nausea-related quality of life (QoL), compared to placebo, in cancer patients undergoing emetogenic chemotherapy	• Double-blind, randomized placebo-controlled trial of 51 cancer patients receiving emetogenic chemotherapy • Patients were fed either 1.2 g of standardized ginger extract or placebo per day, in addition to standard anti-emetic therapy, during the first three cycles of chemotherapy • The acute and delayed nausea, vomiting, cancer-related fatigue, nutritional status, etc., were observed	• Over three consecutive chemotherapy cycles, nausea was more prevalent than vomiting • In Cycle 1, participants reported significantly better QoL, chemotherapy-induced nausea and vomiting	• The adjuvant supplementation of ginger is more related to chemotherapy-induced nausea and cancer-related quality of life, and is less associated with cancer-related fatigue
Tan-Shalaby et al. (2016) NCT01716468	To study the effect of ketogenic diet in advanced cancer across different tumour types	• Recruited 17 advanced cancer patients devoid of chemotherapy	• Significant weight lost was observed with stable profile of haematologic, biochemical	• Modified Atkins diets were found to be safe in case of advanced cancer patient

		• Patients were fed 20–40 g of carbohydrates/day up to 16 weeks • Weight loss was correlated with serum ketones, glucose, lipids and creatinine	and lipids • No correlations between serum glucose, ketones or lipids and clinical response • Two patients developed brain metastases and survived 80 and 116 weeks respectively	• Steroid treatment restored optimal ketone and glucose levels
Zhang et al. (2016) NCT00843167	To examine the relationship between dietary broccoli intake and study of biomarkers for histone acetylation and cell cycle regulation from breast tissue	• A randomized trial of 54 women participants received oral broccoli sprout extract supplementation three times daily for 2–8 weeks • Breast tissue was obtained from biopsy samples, and effect was analysed on biomarkers of prognosis and HDAC inhibition in peripheral blood cell and normal and cancerous breast tissue samples from these patients	• Total cruciferous vegetable intake was inversely associated with Ki-67 protein expression in breast ductal carcinoma in situ (DCIS) tissue • No association was found for other biomarkers measured (HDAC3, HDAC6, H3K9, H3K18 and p21) in all tissues examined (benign, DCIS and IDC)	• The trial provided additional evidence for the potential role of sulforaphane (broccoli sprout extract) in histone acetylation and cell proliferation • Total cruciferous vegetable intake was associated with decreased cell proliferation in breast DCIS tissue
Sidahmed et al. (2016) NCT00475722	To investigate the effect of Mediterranean diet on human subjects with increased risk of colorectal cancer	• A randomized trial of 120 participants with an increased risk for colorectal cancer • A Mediterranean diet or a healthy diet was feed for 6 months • Colon biopsies were obtained before and after 6 months of intervention for prostaglandin pathway gene analysis	• The most highly expressed genes were prostaglandin dehydrogenase (HPGD) and prostaglandin H synthases (PTGS1) • Colon PTGS1 expression and SFA concentrations were identified as important factors affecting inter-individual differences in colon PGE_2 concentrations	• Short-term dietary change had little effect on enzymes in the prostaglandin pathway in the colon both in Mediterranean and healthy diet • Both dietary patterns reduced dietary intakes of saturated fats and had little effect on expression of genes in the PGE2 pathway

(continued)

Table 5.1 (continued)

Author and trial ID	Purpose of clinical trial	Design and description of the trial	Outcome of the trial	Conclusion
Charron et al. (2015) NCT01293591	To probe the mechanism of raw, crushed garlic (RCG) on expression of biomarker genes related to immunity and cancer	• A randomized crossover feeding trial in 17 volunteers with consumption of 5 g RCG and or a garlic-free control meal up to 10 days • Expression profile of selected immunity and cancer-related genes from blood was analysed using qRT-PCR	• RCG induced upregulation of seven genes (AHR, ARNT, HIF1A, JUN, NFAT, NFAM1, OSM and V-rel)	• RCG upregulated genes related to immunity, apoptosis and xenobiotic metabolism in humans
Rieger et al. (2014) NCT00575146	To assess the effect of mild ketogenic diet on quality of life and survival of patients with recurrent glioblastoma (PRG)	• An interventional trial of 20 PRG • Patients were put on a low-carbohydrate, ketogenic diet containing plant oils	• A ketogenic diet is feasible and safe but probably has no significant clinical activity when used as single-agent PRG • Median progression-free survival of all patients was 32 weeks since enrolment	• The unrestricted ketogenic diet was found safe and tolerable by PRG • The ketogenic diet intervention can be safe for PRG
Magbanua et al. (2011) NCT00402285	To study the effect of lycopene and fish oil on prostate gene expression in low-risk prostate cancer volunteers	• A randomized trial of 84 men with low-risk prostate cancer was given 3-month lycopene or fish oil supplementation or placebo • Differential gene expression and pathway analyses	• Global gene expression analysis revealed no significant individual genes linked with high intake of fish or tomato or after 3 months of supplementation with lycopene or fish oil	• Per say there were significant individual genes associated with dietary intake and supplementation of lycopene and fish oil • Exploratory analyses revealed candidate in vivo

		were performed to identify genes and pathways regulated by these micronutrients	• Exploratory pathway analyses of rank-ordered genes showed the regulation of androgen and oestrogen metabolism in men who routinely consumed more fish and tomato as compared to men who have moderate consumption	pathways that may be regulated by these micronutrients
Young et al. (2011) CT01824498	To study the association of dietary fat, eicosanoids and breast cancer risk	• A controlled crossover feeding trial in postmenopausal women using three diets: High fat (HF; 40% energy from fat), low fat (LF; 20% energy from fat) and low fat plus n-3 FA (LFn3; 20% of energy from fat plus 3% of energy from n-3 FA) in 8-week feeding periods	• Plasma phospholipid fatty acid n-3 increased with the LFn3 relative to HF and LF • Plasma oestradiol (51%), urinary prostaglandin E and 11-dehydro-thromboxane B2i increased with inclusion of HF relative to LF • Varying omega-3 (n-3) fatty acid (FA) intake was associated with changes in circulating sex hormones	• Breast cancer risk appears to be increased by the HF and low risk with LF • LFn3 did not result in significant changes in sex hormones • The rate of breast cancer risk increases with the increase in sex hormone concentration
Olendzki et al. (2009) NCT00561977	The aim of the trial was to find a simple message to improve dietary quality for cancer and heart disease	• A random trial of 36 individuals who were provided individual dietary counselling sessions • Patients were divided into three arms: (1) Low saturated fat diet, (2) High-fibre diet and (3) Low saturated fat and high fibre • Compared the feasibility of arms	• Dietary fibre intervention counselling was easier to maintain as compared with the other two intervention conditions • The high-fibre condition decreased the saturated fat intake of participants • Participants in all three counselling conditions lost an average of 9 lb from baseline weight	• A simple dietary message is feasible and can improve overall dietary quality • Overall dietary quality improved in all three arms

(continued)

Table 5.1 (continued)

Author and trial ID	Purpose of clinical trial	Design and description of the trial	Outcome of the trial	Conclusion
Dhillon et al. (2008) Not registered	To evaluate the clinical biological inhibitory effects of curcumin against NF-kB and cytokines in advanced pancreatic cancer patient	• A phase II trial of 25 patients with advanced pancreatic cancer • Patients were fed orally with 8 g curcumin/day up to 18 months • Serum cytokine levels of interleukin IL-6, IL-8, IL-10 and IL-1 receptor antagonists and cell expression of NF-nB in blood cells and COX-2 were analysed	• Circulating curcumin levels were low • No toxicities were observed • Curcumin downregulated expression of NF-KB, COX-2 and phosphorylated signal transducer	• Oral curcumin is well tolerated with limited absorption and has activity against target genes in some patients with advanced pancreatic cancer
Bairati et al. (2006) Not registered	The aim of the trial was to analyse the long-term and high-dose effect of antioxidant vitamins (vitamin E, b-carotene) with all-cause mortality	• The 540 patients with stage I or II head and neck cancer treated by radiation therapy were recruited into supplement arm (a-tocopherol or b-carotene) and placebo arm • The supplements were given during radiation therapy and further up three additional years	• Cause-specific mortality rates were higher in the supplement arm than in the placebo arm • All-cause mortality was significantly increased in the supplement arm	• Long-term supplementation with high-dose antioxidant vitamins is associated with mortality in cancer patient

group (The Alpha-Tocopherol Beta Carotene Cancer Prevention Study Group, 1994). In two large low-fat diet intervention trials, the outcome was contradictory indicating a relation between fat intake with breast cancer reappearance (Pierce et al. 2007; Blackburn and Wang 2007); however, in another clinical study low-fat diet was marginally associated with the reduced risk of breast cancer (Prentice et al. 2006). Overall, the dietary intake of fruit and vegetable has protective role against cancer; however, this has turn out to be illusion in case of thyroid cancer. These findings were identified in EPIC cohort studies, which were designed to investigate the relationship between the intake of fruits, vegetables and fruit juices and differentiated thyroid cancer risk (Zamora-Ros et al. 2018).

Several clinical trials investigating the effect of dietary intervention in different types of cancer are in progress; however, unfortunately, it is difficult to compare the outcomes of different trials for several reasons. The most important reason is that the dietary interventions proposed in the trials are quite heterogeneous in nature (Yao et al. 2013). For example, the term "ketogenic diet" may refer to difference in the composition and quantity of protein: fat ratio (Zilberter and Zilberter 2018). Second important aspect is that the control arms of the registered trials (when exist) usually receive consulting on "healthy dietary habits", which per se is a non-standardized notion, usually not enforced, and is extremely complex to monitor.

Conclusion

The outcome of the clinical trials has remained an imbroglio as there is lack of consistency and convincing clinical evidence from large number of RCTs, observational and cohort studies. Moreover, none of the clinical trials at phase 3 of a dietary supplement or food intervention has demonstrated sufficient efficacy and safety in preventing or treating cancer, and therefore, FDA has not approved any dietary supplement or food either to prevent cancer, arresting its growth or prevent its recurrence (Paller et al. 2016). Even a cursory look at the WCRF-AICR dietary guidelines recommended for different types of cancers, invariably there is lack of strong convincing evidence and dietary guidelines are mostly based on limited suggestive and limited non-conclusive evidence (WCRF (World Cancer Research Fund)/American Institute for Cancer Research. Continues Update Project Expert Report 2018). Almost similar type of evidence lacunae was observed with majority of the completed clinical trials (Table 5.1). Perhaps this might be a possible reason for why the oncologists are challenged in providing a satisfactory dietary advice to patients about the efficacy and safety of foods and supplements for cancer prevention or for treating the side effects of cancer treating-therapeutic approaches. Upcoming clinical settings should customize more on molecular underpinning and design a framework of examining how diet impacts on tumour/cancer cell metabolism with possible mode of action in different cancers (Lien and Vander Heiden 2019).

References

Aglago EK, Huybrechts I, Murphy N, Casagrande C, Nicolas G, Pischon T, et al. Consumption of fish and long-chain n-3 polyunsaturated fatty acids is associated with reduced risk of colorectal cancer in a large European cohort. Clin Gastroenterol Hepatol. 2019. pii: S1542-3565(19)30669-X.

Bairati I, Meyer F, Jobin E, Gélinas M, Fortin A, et al. Antioxidant vitamins supplementation and mortality: a randomized trial in head and neck cancer patients. Int J Cancer. 2006;119(9):2221–4.

Beynon RA, Richmond RC, Santos Ferreira DL, Ness AR, et al. Investigating the effects of lycopene and green tea on the metabolome of men at risk of prostate cancer: the ProDiet randomised controlled trial. Int J Cancer. 2019;144(8):1918–28.

Blackburn GL, Wang KA. Dietary fat reduction and breast cancer outcome: results from the Women's Intervention Nutrition Study (WINS). Am J Clin Nutr. 2007;86:S878–81.

Bondonno NP, Dalgaard F, Kyro C, Murray K, Bondonno CP, et al. Flavonoid intake is associated with lower mortality in the Danish Diet Cancer and Health Cohort. Nat Commun. 2019;10(1):3651.

Charron CS, Dawson HD, Albaugh GP, Solverson PM, Vinyard BT, Solano-Aguilar GI, Molokin A, Novotny JA. A single meal containing raw, crushed garlic influences expression of immunity- and cancer-related genes in whole blood of humans. J Nutr. 2015;145(11):2448–55.

Cohen CW, Fontaine KR, Arend RC, Gower BA. A ketogenic diet is acceptable in women with ovarian and endometrial cancer and has no adverse effects on blood lipids: a randomized, controlled trial. Nutr Cancer. 2019;72(4):584–94. https://doi.org/10.1080/01635581.2019.1645864.

de van der Schueren MAE, Laviano A, Blanchard H, Jourdan M, Arends J, Baracos VE. Systematic review and meta-analysis of the evidence for oral nutritional intervention on nutritional and clinical outcomes during chemo(radio)therapy: current evidence and guidance for design of future trials. Ann Oncol. 2018;29:1141–53.

Dehghan M, Mente A, Zhang X, et al. Associations of fats and carbohydrate intake with cardiovascular disease and mortality in 18 countries from five continents (PURE): a prospective cohort study. Lancet. 2017;390(10107):2050–62.

Dhillon N, Aggarwal BB, Newman RA, Wolff RA, Kunnumakkara AB, et al. Phase II trial of curcumin in patients with advanced pancreatic cancer. Clin Cancer Res. 2008;14(14):4491–9.

Dinu M, Pagliai G, Casini A, Sofi F. Mediterranean diet and multiple health outcomes: an umbrella review of meta-analyses of observational studies and randomised trials. Eur J Clin Nutr. 2018;72(1):30–43. https://apps.who.int/iris/bitstream/handle/10665/275990/9789241514873-eng.pdf?ua=1.

Key TJ, Allen NE, Spencer EA, Travis RC. The effect of diet on risk of cancer. Lancet Lond Engl. 2002;360(9336):861–8.

Lévesque S, Pol JG, Ferrere G, Galluzzi L, Zitvogel L, Kroemer G. Trial watch: dietary interventions for cancer therapy. Oncoimmunology. 2019;8(7):1591878.

Lien EC, Vander Heiden MG. A framework for examining how diet impacts tumour metabolism. Nat Rev Cancer. 2019;19:651–61. https://doi.org/10.1038/s41568-019-0198-5.

Magbanua MJ, Roy R, Sosa EV, Weinberg V, Federman S, Mattie MD, Hughes-Fulford M, Simko J, Shinohara K, Haqq CM, Carroll PR, Chan JM. Gene expression and biological pathways in tissue of men with prostate cancer in a randomized clinical trial of lycopene and fish oil supplementation. PLoS One. 2011;6(9):e24004.

Manson JE, Cook NR, Lee IM, Christen W, Bassuk SS, Mora S, et al. Vitamin d supplements and prevention of cancer and cardiovascular disease. N Engl J Med. 2019;380:33–44.

Marx W, McCarthy AL, Dan McKavanagh KR, Vitetta L, Sali A, Lohning A, Isenring E. The effect of a standardized ginger extract on chemotherapy-induced nausea-related quality of life in patients undergoing moderately or highly emetogenic chemotherapy: a double blind, randomized, placebo controlled trial. Nutrients. 2017;9(867):11–3.

Mayne ST, Ferrucci LM, Cartmel B. Lessons learned from randomized clinical trials of micronutrient supplementation for cancer prevention. Annu Rev Nutr. 2012;32:369–90.

Mayne ST, Playdon MC, Rock CL. Diet, nutrition, and cancer: past, present and future. Nat Rev Clin Oncol. 2016;13:504–15.

Meyskens FL, Szabo E. How should we move the field of chemopreventive agent development forward in a productive manner? Recent Results Cancer Res. 2005;166:113–24.

NCD Risk Factor Collaboration (NCD-RisC). Trends in adult body-mass index in 200 countries from 1975 to 2014: a pooled analysis of 1698 population-based measurement studies with 19·2 million participants. Lancet Lond Engl. 2016;387(10026):1377–96.

Nosrati N, Bakovic M, Paliyath G. Molecular mechanisms and pathways as targets for cancer prevention and progression with dietary compounds. Int J Mol Sci. 2017;18:2050.

Nuñez-Sánchez MA, González-Sarrías A, García-Villalba R, et al. Gene expression changes in colon tissues from colorectal cancer patients following the intake of an ellagitannin-containing pomegranate extract: a randomized clinical trial. J Nutr Biochem. 2017;42:126–33.

Olendzki BC, Ma Y, Schneider KL, Merriam P, Culver AL, Ockene IS, Pagoto S. A simple dietary message to improve dietary quality: results from a pilot investigation. Nutrition. 2009;25 (7–8):736–44.

Paller CJ, Denmeade SR, Carducci MA. Challenges of conducting clinical trials of natural products to combat cancer. Clin Adv Hematol Oncol. 2016;14(6):447–55.

Pierce JP, et al. Influence of a diet very high in vegetables, fruit, and fiber and low in fat on prognosis following treatment for breast cancer: the Women's Healthy Eating and Living (WHEL) randomized trial. JAMA. 2007;298:289–98.

Prentice RL, et al. Low-fat dietary pattern and risk of invasive breast cancer: the Women's Health Initiative Randomized Controlled Dietary Modification Trial. JAMA. 2006;295:629–42.

Rieger J, Bähr O, Maurer GD, Hattingen E, Franz K, Brucker D, Walenta S, Kämmerer U, Coy JF, Weller M, Steinbach JP. ERGO: a pilot study of ketogenic diet in recurrent glioblastoma. Int J Oncol. 2014;44(6):1843–52.

Sidahmed E, Sen A, Ren J, Patel A, Turgeon DK, Ruffin MT, Brenner DE, Djuric Z. Colonic saturated fatty acid concentrations and expression of COX-1, but not diet, predict prostaglandin E2 in normal human colon tissue. Nutr Cancer. 2016;68(7):1192–201.

Singh AK, Sharma N, Ghosh M, Ho YP, Jeong DK. Emerging importance of dietary phytochemicals in fight against cancer: role in targeting cancer stem cells. Crit Rev Food Sci Nutr. 2017;57:16.

Szmulewitz RZ, Peer CJ, Ibraheem A, et al. Prospective international randomized phase II study of low-dose biraterone with food versus standard dose abiraterone in castration-resistant prostate cancer. J Clin Oncol. 2018;36(14):1389–95.

Tan-Shalaby JL, Carrick J, Edinger K, Genovese D, Liman AD, Passero VA, Shah RB. Modified Atkins diet in advanced malignancies—final results of a safety and feasibility trial within the Veterans Affairs Pittsburgh Healthcare System. Nutr Metab (Lond). 2016;13:52. https://doi.org/10.1186/s12986-016-0113-y.

WCRF (World Cancer Research Fund)/American Institute for Cancer Research. Continues Update Project Expert Report. Recommendations and public health and policy implications. 2018. dietandcancerreport.org. Accessed 24 Sept 2020.

World Health Organization WHO Global nutrition policy review 2016–2017. 2018. Accessed 29 Sept 2020.

Yang CS, Chen JX, Wang H, Lim J. Lessons learned from cancer prevention studies with nutrients and non-nutritive dietary constituents. Mol Nutr Food Res. 2016;60(6):1239–50.

Yao CK, Gibson PR, Shepherd SJ. Design of clinical trials evaluating dietary interventions in patients with functional gastrointestinal disorders. Am J Gastroenterol. 2013;108(5):748–58.

Young LR, Kurzer MS, Thomas W, Redmon JB, Raatz SK. Effect of dietary fat and omega-3 fatty acids on urinary eicosanoids and sex hormone concentrations in postmenopausal women: a randomized controlled feeding trial. Nutr Cancer. 2011;63(6):930–9.

Zamora-Ros R, Béraud V, Franceschi S, et al. Consumption of fruits, vegetables and fruit juices and differentiated thyroid carcinoma risk in the European Prospective Investigation into Cancer and Nutrition (EPIC) study. Int J Cancer. 2018;142(3):449–59.

Zhang Z, Atwell LL, Farris PE, Ho E, Shannon J. Associations between cruciferous vegetable intake and selected biomarkers among women scheduled for breast biopsies. Public Health Nutr. 2016;19(7):1288–95.

Zilberter T, Zilberter Y. Ketogenic ratio determines metabolic effects of macronutrients and prevents interpretive bias. Front Nutr. 2018;5:75.

Chapter 6
Diet and Cancer Epigenetics: A Step Towards Developing Cancer Epigenetic Diet

Abstract The remarkable breakthrough research in molecular biology and advancements in next-generation sequencing (NGS) have significantly enriched our understandings of the epigenetic regulation of gene expression. The advancement in epigenetic research also inspired the scientific interest for investigating the influence of diet on variety of degenerative diseases including cancer. However, the molecular underpinnings underlying these effects are poorly understood. The scientific enquiry as to how dietary ingredients influence epigenetic modifications that eventually regulate health was attributed with catalytic functions of the epigenetic enzymes, which are responsible for "writing" or "erasing" the epigenetic modifications. Voluminous preclinical data suggest that dietary bioactive ingredients modulate the epigenetic landscape by inhibiting/regulating the activity of HDACs, HAT, DNMTs, acetylation/deacetylation, sumyolation, chromatin remodelling, etc. These diet-induced histone modifications regulate the conformational dynamics of chromatin structure, which ultimately leads to changes in DNA accessibility by transcription machinery leading to gene expression. Variety of dietary bioactive molecules found in several foods act as dietary modulators, which alter the epigenetic landscape against the growth and proliferation of cancer. Searching such diet origin anticancer epigenetic modulators and making a consortium of such dietary sources will lead to the formulation of anticancer epigenetic diet.

Keywords Diet · Nutrition · Epigenetics · HADC · Histones · DNMT · Regulation · Epigenetic diet

Abbreviations

BRCA	Breast cancer growth suppressor protein or BReast CAncer gene
DAPK	Death-associated protein kinase
DDX	DEAD-box helicase
DGCR8	DGCR8 microprocessor complex subunit
DNA	Deoxyribonucleic acid
DNMTs	DNA methyltransferases

R. N. Gacche, *Dietary Research and Cancer*,
https://doi.org/10.1007/978-981-16-6050-4_6

EGCG	Epigallocatechin-3-gallate
ER	Oestrogen receptor, endoplasmic reticulum
HAT	Histone acetyltransferase
HDAC	Histone deacetylase
HMC	Hallmark of cancer
HORMAD1	HORMA domain-containing protein 1
MAEL	Maelstrom spermatogenic transposon silencer
MAGEA11	MAGE family member A11
MAGEA2	MAGE family member A2
MAGEB2	MAGE family member B2
MAGEC2	MAGE family member C2
NGS	Next-generation sequencing
NXF2	Nuclear RNA export factor 2
PIWIL1	Piwi-like RNA-mediated gene silencing 1
pRB	Phosphorylated retinoblastoma protein
RARβ	Retinoic acid receptor beta
RUNX	Runt-related transcription factor
SSX	Synovial sarcoma, X
SUMO	Small ubiquitin-like modifier
TGF-β	Transforming growth factor β
TP53	Tumour protein p53

Introduction

The advent of technologies like next-generation sequencing (NGS) gave a strong impetus towards molecular understanding of cancer pathophysiology and progression. In fact, it was NGS technology since past decades, which unravelled the genomic landscapes of common forms of human cancer, which otherwise was considered as a science fiction. For majority of the cancer types, the cancer genome landscape consists of ~140 genes, when altered by intragenic mutations can "drive" tumorigenesis (driver mutations). Typically, a tumour has two to eight of such "driver gene" mutations, while remaining mutations are passengers which as such do not contribute towards selective growth advantage to growing tumours. Further, the driver genes can be categorized into 12 signalling pathways that coordinate the regulation of three core cellular processes like cell survival, cell fate and genome maintenance (Vogelstein et al. 2013). The cancer genome landscapes have opened new avenues of understanding the pathophysiology of cancer signalling pathways, which perhaps was one of the most pressing needs in the mainstream of basic cancer research. The new clinical settings of dietary interventions should rationalize the findings of cancer genome and probe the possible interactive cancer drug targets, which selectively inhibit the cancer growth. To a greater extent, the evolving current literature of diet–cancer studies has taken a cognizance of cancer genome landscapes for understanding the mechanistic anticancer potential of diet and nutrition (Nosrati et al. 2017).

Another important conceptual scientific progress that needs to articulate for understanding the diet–cancer interactions is the "hallmarks of cancer" (HMC). HMC is the eight (two added to existing six) biological characteristics acquired during the multistep process of human cancer progression. The eight HMC comprises of sustained proliferative signalling, combating growth suppressors, circumventing cell death, enabling replicative immortality, inducing angiogenesis, activating invasion and metastasis, reprogramming of energy metabolism and evading immune destruction (Hanahan and Weinberg 2011). There lies an opportunity to investigate the molecular underpinnings of dietary ingredients, which counteract the HMC and integrate it in clinical trials so as to generate the strong convincing evidence of dietary intervention for cancer prevention.

Diet and Cancer Epigenetics: A Step Towards Developing Cancer Epigenetic Diet

Epigenetics deals with the heritable modifications at genomic level that usually do not alter the DNA sequence. Epigenetic changes are manifested through series of processes like DNA methylation, histone modification (acetylation/deacetylation), sumyolation, chromatin remodelling and miRNA-regulated gene expressions. The set of enzymes and other factors in a very circumstantial situation coordinate the decision of gene expression or suppression. Epigenetic modifications have profound influence on human behaviour, appearance, disease susceptibility, stress response and even ageing. Per say, epigenetic gene regulation is instrumental in regulating normal physiological processes and also equally implicated in the pathophysiology of several human degenerative diseases including cancer. On one hand, epigenetic mechanisms buffer the environmental variation; on the other hand, it allows plastic responses to the most extreme environmental cues (Cavalli and Heard 2019).

Dysregulation of the epigenome has been identified as a hallmark of human cancers and has been closely associated with tumour pathogenesis and progression. Since epigenetic modifications are reversible, the experimental settings are being designed to identify the impact of diet on the landscape of epigenome with a hope of developing a strategic consortium of dietary ingredients, which may induce epigenetic reprogramming and maintain the sequel of normal gene expression profile unlike that of aberrant cancer cell signalling (Nasir et al. 2020). In recent years, diet and epigenetic modification in relation to role of dietary bioactive ingredients in modifying the epigenetic pathways that offer benefits by neutralizing the aberrant gene expression profile of cancer patient has attracted the scientific community (Lewis and Tollefsbol 2017). Substantial scientific evidence is accumulating in recent years linking the role of commonly used bioactive dietary metabolites as influential agents in modifying the landscape of epigenome (Montgomery and Srinivasan 2019).

DNA methylation (a reaction catalysed by DNA methyltransferases leading to methylation of fifth carbon of cytosine at CpG sites) patterns is one of the most extensively studied epigenetic modifications in mammalian cancers. Aberrant methylation patterns of genes regulating the cell cycle, apoptosis, proliferation, metastasis and drug resistance have been identified in different cancer types. The excitement of developing the inhibitors of DNA methyltransferases (DNMTs) as anticancer agents pop up, when several findings revealed that the hypermethylation of the promoters of series of tumour suppressor genes such as TP53, APC, p14, p15, p21WAF1/CIP1, p16, MLH1, RUNX3, TIG1, 1A (RASSF1A), BRCA1, pRB, TGF-β, MGMT, RARβ2, DAPK, oestrogen receptor (ER) alpha and progesterone receptor in different cancers are hypermethylated, which may not get expressed and may contribute towards genetic instability and cancer development. An extensive review describing the degree of methylation (%) of promoters of above described tumour suppressor genes has been summarized elsewhere, and the reports are accumulating in logarithmic fashion (Herman and Baylin 2003; Baylin and Ohm 2006; Esteller 2008; Wang et al. 2018). As hypermethylation suppresses the tumour suppressor genes, inversely, hypomethylation is associated with the activation of certain oncogenes implicated in tumorigenesis. Series of aberrant "cancer germline" genes such as SSX, DDX43, DDX53, MAEL, BORIS/CTCFL, PIWIL1), CT-GABRA3, MAGEA11, MAGEB2, MAGEA2, MAGEC2, HORMAD1, NXF2 and MAGEA3/MAGEA6 are activated in variety of oncogenic pathways owing to their hypomethylation (reviewed by Van Tongelen et al. 2017).

In the quest of searching novel inhibitors of DNMTs as cancer therapeutic agents, variety of dietary sources and their bioactive metabolites such as parsley/celery/apple/wheat sprouts/red onions/oranges (apigenin and luteolin), citrus/dark cherries/berries (quercetin), citrus (naringenin and hesperidin), berries (myricetin, ellagic acid), red grapes (syringic acid), beetroot (betanin), berries/grapes (cyaniding), apple (phloretin), mustard (synaptic acid), cinnamon (coumaric acid), cruciferous vegetables (sulforaphane), coffee (caffeic acid and chlorogenic acid), grapes/blueberries (resveratrol, piceatannol), soya bean (genistein, biochanin A, daidzein), green tea (catechin, epicatechin, epicatechin gallate, epigallocatechin, epigallocatechin-3-gallate), turmeric (curcumin) and olives/white wine grapes/onions/mushrooms (protocatechuic acid) have demonstrated significant inhibition of DNMTs in preclinical model studies (Fig. 6.1). Few of the bioactive molecules have also been tested in clinical trials, but not with a focus on DNMT inhibition (Link et al. 2010; Vahid et al. 2015; Rajnee et al. 2016; Andreescu et al. 2018; Nasir et al. 2020).

Another important epigenetic mechanism regulating the gene expression is through histone modifications. The dynamics of gene expression or repression is closely associated with the histone modifications via methylation, acetylation (at the N-terminal group of lysine residues), deacetylation, ubiquitylation, phosphorylation, sumoylation, biotinylation, ADP ribosylation, etc. Among these modifying reactions, the histone acetylation, deacetylation and methylation are the most extensively

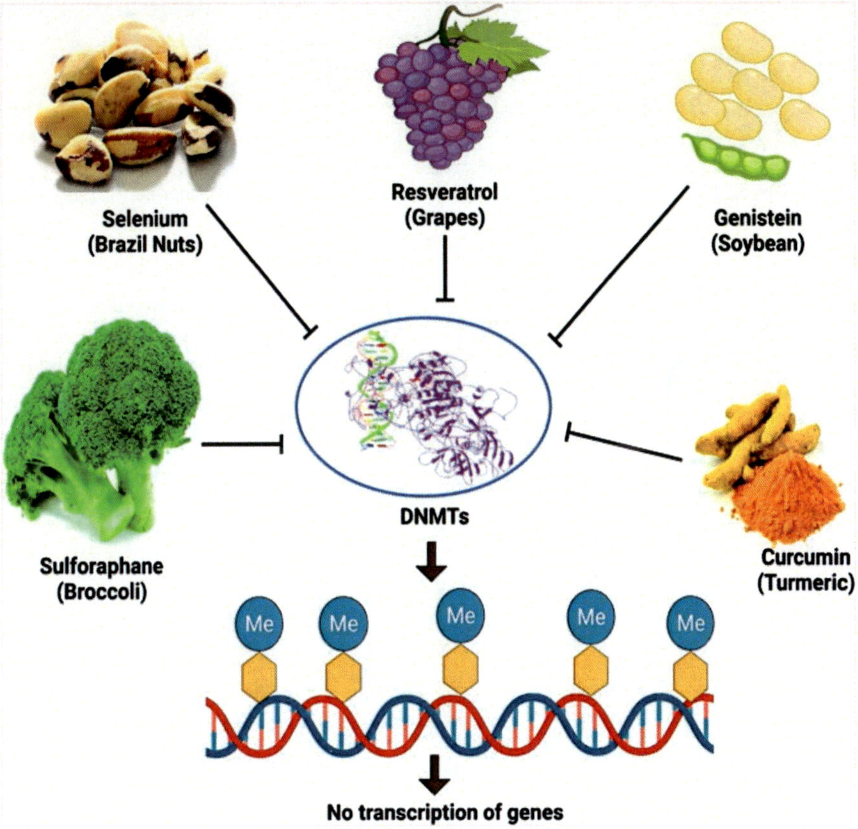

Fig. 6.1 Representative dietary inhibitors of DNA methyltransferases (DNMTs). The DNMT enzyme catalyses the process of DNA methylation by adding a methyl group (CH3; shown by Me) onto cytosines residue of CpG dinucleotides (benzene ring). Hypermethylation of CpG dinucleotides or CpG islands by DNMTs usually leads to transcriptional gene silencing and gene inactivation. Variety of dietary bioactive molecules found in several foods act as dietary inhibitors of DNMTs and thereby alter gene expression via epigenetic mechanisms

investigated with an intent of developing therapeutic targets. If a gene has to express, the histone–DNA interaction needs to loosen so as to provide access to transcription assembly. This is achieved via acetylation of the histone, wherein the lysine residue is acetylated by histone acetyltransferase (HAT) and removes its positive charge leading to repulsion between histone and DNA. This open conformation of chromatin can be accessed by various transcription factors and promotes the transcriptional activation of genes (Audia and Campbell 2016). Conversely, deacetylation of lysine residue by histone deacetylase (HDAC) restores the positive charge of lysine and strengthens the histone–DNA interaction, thereby not providing access to transcriptional machinery and repressing the gene expression. Of note, post-transcriptional modifications of histone are dysregulated by dysregulation of HDACs in variety of cancers and there is increased activity profile of HDACs in different cancer types.

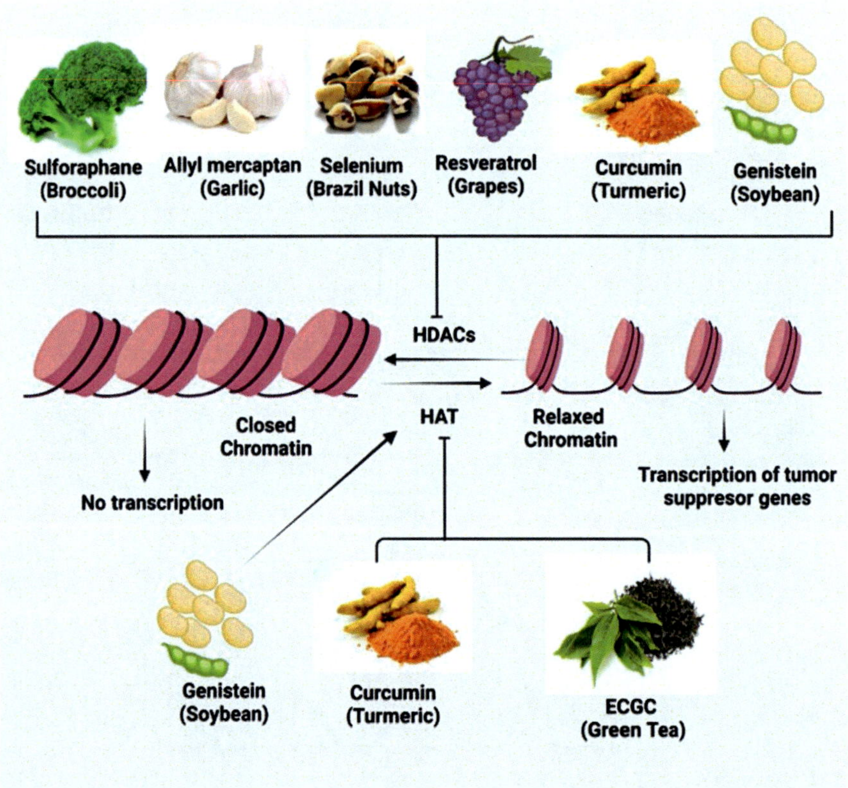

Fig. 6.2 Dietary bioactive ingredients modulate the epigenetic landscape by inhibiting/regulating the activity of HDAC and HAT. These histone modifications regulate the conformational dynamics of chromatin structure, which ultimately leads to changes in DNA accessibility by transcription machinery. HATs catalyse the acetylation of histones leading to relaxed chromatin conformation, which allows transcriptional factors to access DNA and activates gene expression; on the other hand, chromatin with its closed configuration represents gene silencing and repression mode of gene expression. This state of gene silence is maintained by deacetylation of histones by HDACs. Therefore, the dietary ingredients inhibiting the HDAC activity are important for designing diet for cancer patients

Nevertheless, recent literature links the functions of HDACs in cancer initiation, progression, metastasis, etc., and provides cues of developing crucial attractive target against cancer (Patra et al. 2019).

Plethora of preclinical studies has accumulated in the recent past linking the role of dietary metabolites as HDAC inhibitors (Fig. 6.2). Variety of commonly consumed dietary sources and identified bioactive molecules like broccoli (sulforaphane, isothiocyanates, 3,3-diindolylmethane), ginger (6-Shogaol, zerumbone, 6-methoxy-2E,9E-humuladien-8-one), garlic (*S*-allylmercaptocysteine, diallyl disulphide and trisulphide, allyl mercaptan), soya (biochanin A, daidzein, equol, genistein), coffee (caffeic acid, chlorogenic acid), grapes (resveratrol), feijoa (flavone), black and green tea (epigallocatechin-3-gallate, theophylline, polyphenon B), parsley/celery/apple (apigenin), carrots, mushrooms (chrysin), turmeric (curcumin), dark cherries/berries/citrus/broccoli/red raspberry (quercetin) and pomegranates/raspberries/walnuts (ellagitannins) have demonstrated significant HDAC inhibition under various preclinical model studies (Link et al. 2010; Rajendran et al. 2011; Vahid et al. 2015; Rajnee et al. 2016; Zam and Khadour 2017; Andreescu et al. 2018; Nasir et al. 2020; Merarchi et al. 2019).

A step ahead, under clinical trial settings (NCT01543074) involving the human subjects, the broccoli sprout extract supplement of 200 μmol sulforaphane equivalents or more than five cruciferous vegetable servings per week resulted in increased p16 expression, and interestingly, it was inversely associated with HDAC3 in circulating peripheral blood mononuclear cells and in biopsies, demonstrating the efficacy of sulforaphane or cruciferous vegetable as HDAC inhibitor in human colon (Rajendran et al. 2015).

Identification and development of novel DNMT and HDAC inhibitors from dietary sources might serve as adjuvant therapy owing to their minimum or no toxicity and increased bioavailability. The quest of searching dietary DNMT and HDAC inhibitors is in tune with developing adjuvant therapeutic approach for cancer prevention, as FDA has already approved vorinostat, romidepsin, belinostat and panobinostat as HDAC inhibitors for the treatment of cutaneous T cell and refractory peripheral T-cell lymphoma (Andreescu et al. 2018) and DNMT inhibitors (decitabine and azacitidine) for the treatment of therapeutic applications of myelodysplastic syndromes. Many times preclinical studies have its own limitations and may not be a clinical reality.

Conclusion

Developing different dietary patterns for the prevention of cancer has remained one of the important thrust areas in the mainstream of diet–cancer research. The opportunity lies in making excursion and bioprospecting the natural resources for the identification of novel, safe and effective anticancer epigenetic modulators (especially DNMT and HDAC inhibitors) from natural sources and validating them clinically under defined set of parameters with specific type of cancer pathophysiology. The conceptual formulation and implementation of "epigenetic diet" lies in the fact that environmental factors have remarkable impact on the functioning of epigenome and that these changes can also be introduced through dietary interventions. In fact, the term "epigenetic diet" was coined to encourage the consumption of specific foods like soya, grapes, cruciferous vegetables, green tea and many others, which have demonstrated epigenetic mechanisms that are against the cancer progression (Hardy and Tollefsbol 2011). There is also an opportunity to investigate metabolically generated dietary intermediates, which possess potential for hijacking histone residues normally modified by more conventional epigenetic processes.

References

Andreescu N, Puiu M, Niculescu M. Effects of dietary nutrients on epigenetic changes in cancer. Methods Mol Biol. 2018;1856:121–39.
Audia JE, Campbell RM. Histone modifications and cancer. Cold Spring Harb Perspect Biol. 2016;8:a019521.
Baylin SB, Ohm JE. Epigenetic gene silencing in cancer—a mechanism for early oncogenic pathway addiction? Nat Rev Cancer. 2006;6:107–16.
Cavalli G, Heard E. Advances in epigenetics link genetics to the environment and disease. Nature. 2019;571(7766):489–99.
Esteller M. Epigenetics in cancer. N Engl J Med. 2008;358:1148–59.
Hanahan D, Weinberg RA. Hallmarks of cancer: the next generation. Cell. 2011;144:646–74.
Hardy TM, Tollefsbol TO. Epigenetic diet: impact on the epigenome and cancer. Epigenomics. 2011;3(4):503–18.
Herman JG, Baylin SB. Gene silencing in cancer in association with promoter hypermethylation. N Engl J Med. 2003;349(21):2042–54.
Lewis KA, Tollefsbol TO. The influence of an epigenetics diet on the cancer epigenome. Epigenomics. 2017;9(9):1–3.
Link A, Balaguer F, Goel A. Cancer chemoprevention by dietary polyphenols: promising role for epigenetics. Biochem Pharmacol. 2010;80(12):1771–92.
Merarchi M, Sethi G, Shanmugam MK, Fan L, Arfuso F, Ahn KS. Role of natural products in modulating histone deacetylases in cancer. Molecules. 2019;24(6). pii: E1047.
Montgomery M, Srinivasan A. Epigenetic gene regulation by dietary compounds in cancer prevention. Adv Nutr. 2019;10(6):1012–28. https://doi.org/10.1093/advances/nmz046.
Nasir A, Bullo MMH, Ahmed Z, Imtiaz A, Yaqoob E, Jadoon M, et al. Nutrigenomics: epigenetics and cancer prevention: a comprehensive review. Crit Rev Food Sci Nutr. 2020;60(8):1375–87.
Nosrati N, Bakovic M, Paliyath G. Molecular mechanisms and pathways as targets for cancer prevention and progression with dietary compounds. Int J Mol Sci. 2017;18:2050.
Patra S, Panigrahi DP, Praharaj PP, et al. Dysregulation of histone deacetylases in carcinogenesis and tumor progression: a possible link to apoptosis and autophagy. Cell Mol Life Sci. 2019;76 (17):3263–82.
Rajendran P, Ho E, Williams DE, Dashwood RH. Dietary phytochemicals, HDAC inhibition, and DNA damage/repair defects in cancer cells. Clin Epigenetics. 2011;3(1):4.
Rajendran P, Dashwood WM, Li L, Kang Y, Kim E, Johnson G, et al. Nrf2 status affects tumor growth, HDAC3 gene promoter associations, and the response to sulforaphane in the colon. Clin Epigenetics. 2015;7:102.
Rajnee ES, Candamo KM, Gupta S. Dietary phytochemicals as epigenetic modifiers in cancer: promise and challenges. Semin Cancer Biol. 2016;40(41):82–99.
Vahid F, Zand H, Nosrat-Mirshekarlou E, Najafi R, Hekmatdoost A. The role dietary of bioactive compounds on the regulation of histone acetylases and deacetylases: a review. Gene. 2015;562 (1):8–15.
Van Tongelen A, Loriot A, De Smet C. Oncogenic roles of DNA hypomethylation through the activation of cancer-germline genes. Cancer Lett. 2017;396:130–7.
Vogelstein B, Papadopoulos N, Velculescu VE, Zhou S, Diaz JLA, Kinzler KW. Cancer genome landscapes. Science. 2013;339(6127):1546–58.
Wang LH, Wu CF, Rajasekaran N, Shin YK. Loss of tumor suppressor gene function in human cancer: an overview. Cell Physiol Biochem. 2018;51:2647–93.
Zam W, Khadour A. Impact of phytochemicals and dietary patterns on epigenome and cancer. Nutr Cancer. 2017;69(2):184–200.

Chapter 7
Role of Dietary Ingredients on Expression of Oncogenic and Tumour Suppressor miRNA

Abstract Approximately 50–85% of the human genome is transcribed, and only 1–3% of cellular transcription corresponds with protein-coding DNA sequences. The NGS showed that majority of the human transcriptome over 97–99% is composed of small noncoding RNAs (ncRNAs). MicroRNAs (miRNAs) are a class of ncRNAs that play an important function in regulating gene expression. The advancements in high-throughput functional analysis have unravelled the regulatory functions of variety of miRNAs, which includes the epigenetic regulators of cancer progression, proliferation, metastasis, cell differentiation and cancer stem cells, and also implicated in the regulation of epithelial-to-mesenchymal transition (EMT: a key factor for metastasis) of cancer cells. The therapeutic and pathophysiological roles of miRNAs have inspired the research community, which can be evidenced by citing registration (until June 2021) of 368 clinical trials (ClinicalTrials.gov) associated with various aspects of miRNAs in human ailments. Interestingly, over 132 clinical trials have been completed involving the intervention of "miRNAs" in human disease biology. Literature accumulated in the recent past demonstrated the role of dietary metabolites as downregulators of oncogenic miRNA and inducing the expressions of tumour suppressor miRNAs. Variety of commonly consumed foods like grapes, onions, garlic, cruciferous vegetables (mostly broccoli), soya bean, green tea, Mentha species, variety of citrus fruits, ginger, turmeric, cabbage, collards, cauliflower, kale and mustard greens have demonstrated the potential to inhibit the oncogenic miRNAs and activate the tumour suppressor miRNAs in variety of cancer under preclinical settings.

Keywords Dietary ingredients · Oncogene · Tumour suppressor · miRNAs · Regulation

Abbreviations

3′-UTR	Three prime untranslated region
5′-UTR	Five prime untranslated region
AGO	Argonaute protein

R. N. Gacche, *Dietary Research and Cancer*,
https://doi.org/10.1007/978-981-16-6050-4_7

AKT	Protein kinase B
DBM	Dietary bioactive molecules
DICER1	Dicer 1, ribonuclease III
DROSHA	Drosha ribonuclease
DS	Dietary substances
dsRNA	Double-stranded RNA
EMT	Epithelial-to-mesenchymal transition
Erk	Extracellular signal-regulated kinase
MAPK	Mitogen-activated protein kinase
miRNA	MicroRNA
mTOR	Mammalian target of rapamycin
Src	Rous sarcoma oncogene cellular homolog
TGF-β	Transforming growth factor-β
TRBP	Transactivation response element RNA-binding protein

Introduction

Influence of Diet on Expression of Oncogenic and Tumour Suppressor miRNA

The outcome of human genome project has fuelled biomedical research and offered the deeper understandings towards the role of miRNAs as epigenetic regulators, prognostic and diagnostic biomarker and a future hope of therapeutic modality. MicroRNAs (miRNAs) are highly conserved small nonprotein-coding RNAs measuring approximately 18–22 nucleotides in length. One of the most prominent mechanisms of miRNA gene regulation is by binding to the 3′ untranslated region of mRNA transcript, which ultimately results in complementary base pairing between miRNA and mRNA leading to cleavage of the primary mRNA transcript followed by mRNA degradation and translational arrest. Insights into the molecular underpinnings of miRNAs in disease development in general and cancer in particular have made miRNAs as an attractive diagnostic or prognostic tools and targets for designing novel therapeutic approaches. When an online search with "microRNA and cancer" keywords was carried out, until end of October 2019, there were over 80 clinical trials registered at Clinicaltrials.gov, out of which 26 studies were found to be completed. Most of the studies are related to validating the miRNA as prognostic or diagnostic biomarker of FDA-approved drug efficacy in specific cancer type. The clinical trial settings with "miRNA, diet and Cancer" are yet to be evolved. Owing to FDA approval for first siRNA drug, Patisiran, for the treatment of a rare genetic polyneuropathy in 2018, the miRNA-based therapeutics is inspired with a hope of designing novel miRNA-based drugs in the new era of cancer therapeutics (Rupaimoole and Slack 2017).

With the advancements in NGS technology, series of research findings have described the functional roles of miRNAs as epigenetic regulators of cancer progression, proliferation, metastasis, cell differentiation and cancer stem cells, and also implicated in the regulation of epithelial-to-mesenchymal transition (EMT: a key factor for metastasis) of cancer cells. More precisely, the miRNAs like miR-10b, miR-17, miR-20a, miR-21, miR-19b-1, miR-18a, miR-19a, miR-27a, miR-92-1, miR-155,miR-373, miR-4y, 24-5p, miR-520c and many more have been identified as oncogenic miRNAs, which are overexpressed in variety of cancers; however, the panel of miRNAs like miR-15, miR-16, let-7, miR-34, miR-48, miR-84, miR-241, miR-15a, miR-16-1, miR-126, miR-150, miR-183, miR-203, miR-206, miR-335, miR-495 and many more have been identified as tumour suppressor miRNAs (Del Carmen Martínez-Jiménez et al. 2018; Gezici and Sekeroglu 2017).

Diet and miRNA Regulation: Suppressing the Oncogenic and Upregulating the Tumour-Suppressing miRNAs

Sizable volume of literature has accumulated in the recent time describing the efficacy of dietary metabolites as downregulators of oncogenic miRNA and inducing the expressions of tumour suppressor miRNAs. Variety of commonly consumed foods like grapes, onions, garlic, cruciferous vegetables (mostly broccoli), soya bean, green tea, *Mentha* species, variety of citrus fruits, ginger, turmeric, cabbage, collards, cauliflower, kale and mustard greens have demonstrated the potential to inhibit the oncogenic miRNAs and activate the tumour suppressor miRNAs in variety of cancer under preclinical settings. Mechanistically, the oncogenic and tumour-suppressing effects of miRNAs are mostly mediated by affecting the cell proliferation and apoptosis-related pathways (PI3K-Akt-mTOR, Akt and ERK signalling, AR, Wnt signalling, p53, Src/Ras/ERK, Erb, MAPK and TGF-β signalling), EMT, invasiveness, migration, metastasis, factors of chemo- or radioresistance, adaption to hypoxia and angiogenesis (Del Carmen Martínez-Jiménez et al. 2018; Gezici and Sekeroglu 2017; Biersack 2016a, b; Devi et al. 2017; Samec et al. 2019). The molecular mechanisms of action of dietary substances (DS) in regulating the functions of oncogenic/tumour suppressor miRNA are poorly understood; however, the recent research has identified possible direct and indirect interaction sites of DS, which regulate the miRNA expression and function. The most important sites include interaction of DS with promoters, transcription factors, Drosha/DGCR8 complex and DISER/TRBP/AGO1-4 complex (Fig. 7.1).

More precisely, the representative dietary bioactive molecules like resveratrol (grapes/berries) activate the tumour suppressor miRNAs (miR-34a/c, miR-122-5p, miR-141, miR-194, miR-200c, miR-299-5p, miR-338-3p, miR-582-3p, miR-622, miR-663, miR-758, miR-774) and oncogenic miRNAs (miR-17, miR-21, miR-25, miR-27a, miR-92a-2, miR-103-1, miR-103-2, miR-183, miR-200-3p, miR-520h, miR-542-3p, miR-17-92) in different cancers like colorectal, lung, pancreas, prostate

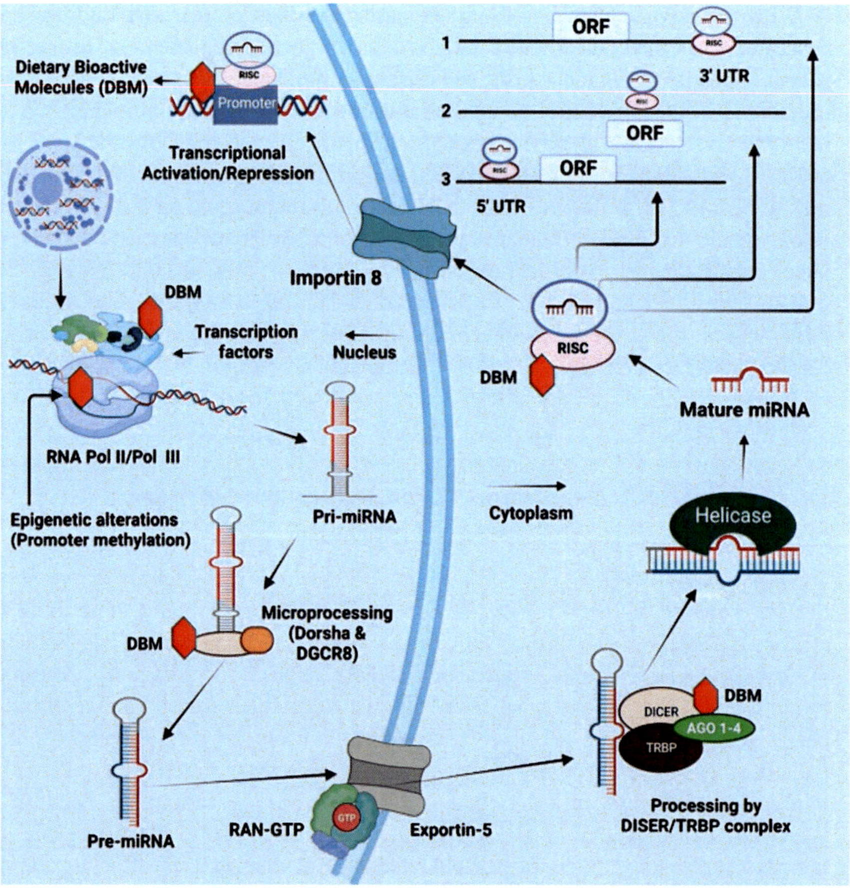

Fig. 7.1 Outline of biogenesis and functions of miRNAs in cellular system. Classically, in the canonical way, pri-miRs are originally transcribed from introns of the protein-coding genes. Further, pri-miRs are subjected for cleavage carried out by the RNase III type endonuclease called DROSHA along with its associated partner proteins called Drosha complex. The processing by Drosha complex leads to generation of a 70-nt stem-loop containing pre-miRNAs. Once the pre-miRNAs are exported from the nucleus to the cytoplasm, it is acted by DICER1 (another RNase III type endonuclease) and generates 22-nt mature phenotype of miRNAs. AS a final step of biogenesis, one of the double-strand mature miRNAs is selectively loaded onto the RISC complex, which comprises of AGOs (1–4) as a core protein component. As a part of functional aspect, the mature miRNAs mostly bind to the 3′-UTR region of the target mRNA, and silence its expression by both mRNA destabilization and translational repression. A subset of miRNAs also binds to the 5′-UTR and ORF of the target mRNA, and thereby repress or activate its translational potential. The dietary bioactive molecules (DBMs) alter the expression of miRNAs by interacting with promoters, transcription factors, Drosha/DGCR8 complex and DISER/TRBP/AGO1-4 complex

and gastric cancers (Wang et al. 2019). The antioxidant flavonoid quercetin found in onions, apple, grapes, citrus species, etc., upregulates the tumour suppressor miRNAs such as let-7, miR-16, miR-26, miR-34a, miR-142-3p and miR-146a and oncogenic miRNAs like miR-27a and miR-125b-3p in colon, lung and pancreas cancers (Akbari et al. 2019). Genistein and soya isoflavones (soya bean) upregulate the tumour suppressor miRNAs such as let-7, miR-23b, miR-34a, miR-146a, miR-200, miR-574-3p and miR-1296 and downregulate different oncogenic miRNAs like miR-21, miR-23b, miR-27a, miR-151, miR-155, miR-221, miR-222, miR-223 and miR-1260b in renal, ovarian, prostate and pancreas cancers (Devi et al. 2017). While curcumin upregulates the panel of tumour suppressor miR-7, miR-9, miR-15, miR-16, miR-22, miR-29a, miR-145, miR-181, miR-192-5p, miR-200b/c, miR-203, miR-205-5p, miR-215 and let-7 and download regulates the oncogenic miRNAs like miR-17-5p, miR-20a, miR-21, miR-27a, miR-186 and miR-208 in breast, colorectal, prostate, lung cancers, leukaemia and hepatocellular carcinoma (Momtazi et al. 2016). Other variety of dietary biometabolites such as apigenin, green tea polyphenols (catechin, epigallocatechin gallate), indole-3 carbinol (Brassica vegetable) and sulforaphane (cruciferous vegetables) inhibit the oncogenic miRNAs and upregulate the tumour suppressor miRNAa by affecting series cell signalling pathways and other factors related to cancer progression (Del Carmen Martínez-Jiménez et al. 2018; Gezici and Sekeroglu 2017; Biersack 2016a, b; Devi et al. 2017; Samec et al. 2019). The above literature describing the role of dietary ingredients in regulation of miRNAs implicated in cancer progression has also been presented in Table 7.1.

Although the miRNA-based therapeutics is yet to reach at the doorstep of clinic, the novel and exciting perspectives of their therapeutic applications are being anticipated and investigated. For example, based on the preclinical and clinical observations it has been speculated that a pool of diet-derived miRNAs may be associated with recipient's health. It is well-known fact that plant genome has large number of miRNA-encoding genes, and these miRNAs are stable owing to the presence of 2′-*O*-methylated terminal nucleotide, which makes them resistant to oxidation, thereby maintaining their bioavailability (Chen 2005). A clinical observation study showed a positive correlation between the amount of plant miRNAs in human blood and the breast cancer prognosis. The study group also demonstrated in vitro cell growth inhibition and in vivo tumour growth inhibition in breast cancer cells using the plant-derived miR-159 (Chin et al. 2016). In a randomized dose-controlled trial of large nutrigenomics cohort study, a positive correlation between the amount of broccoli consumption and the amount of circulating microRNAs in the blood was observed. The authors also demonstrated that *Brassica*-derived microRNAs were capable of regulating the cancer-related gene expression of human genes under in vitro settings and that the microRNAs were found to cooperate with other Brassica-specific compounds in a possible cancer-preventive mechanism (Pastrello et al. 2016). Based on the circumstantial literature describing the role of dietary miRNAs, a "dietary xenomiR hypothesis" has been proposed, which states that miRNAs in foodstuffs can sustain transit through the mammalian GI tract, pass without any modification into host cells and interact with endogenous RNAs to regulate their expression profile in epigenetic manner (Philip et al. 2015). Following this exciting functional and hypothetical dimension of dietary miRNAs as possible anticancer therapeutic strategy, there are many controversial reports accumulated

Table 7.1 Summary of effect of dietary metabolites on regulation of cancer-associated miRNAs

	Effect on miRNA regulation	
Dietary metabolites	Upregulation of tumour suppressor miRNAs	Downregulation of oncogenic miRNAs
Curcumin	miR-7, miR-9, miR-15, miR-16, miR-22, miR-29a, miR-145, miR-181, miR-215, let-7, miR-130a, miR-378, miR-146a, miR-15a, miR-192-5p, miR-200b/c, miR-203, miR-205-5p	miR-27a, miR-208, miR-182, miR-7641, miR-199a, miR-34a, miR-183-96, miR-186, miR-22, miR-17-5p, miR-20a, miR-21
EGCG	miR-1202, miR-1225-5p, miR-1915, miR-let-7-b, let-7a, miR-16, miR-221, miR-2861, miR-3196, miR-34a, miR-3656, miR-3665, miR-205-3p	miR-210, miR-3162, miR-3196, miR-205-3p, miR-16, miR-98-5p, miR-1207-5p, miR-1973, miR-let-7-b, miR-18a, miR-34b, miR-193b, miR-222, miR-342
Sulforaphane	miR-146, miR-23b, miR-381, miR-140, miR-29a	miR-27b, miR-23, miR-92b, miR-221, miR-21, miR-616, miR-155
Resveratrol	miR-125b, miR-409, miR-542, miR-129, miR-204, miR-489, miR-150, miR-296–5p, miR-663, miR-744, miR-622, miR-34a/c, miR-122-5p, miR-141, miR-194, miR-200c, miR-299-5p, miR-338-3p, miR-582-3p, miR-758, miR-774	miR-25, miR-27a, miR-92a-2, miR-20a, miR-18b, miR-19, miR-7, miR-20b, miR-92b, miR-106a, miR106b miR-21, miR-17-92, miR-17, miR-103-1, miR-103-2, miR-183, miR-200-3p, miR-520h, miR-542-3p, miR-424, miR-503, miR-30a-5p
Lycopene	miR-let-lf-1, miR-199a/b, miR214	–
Quercetin	let-7, miR-26, miR-34a, miR-155, miR-146a, miR-22, miR-16, miR-142-3p, miR-146a	miR-21, miR-27a, miR-148, miR-125b-3p
Myricetin	miR-146b	miR-489
Luteolin	miR-21, miR-203, miR-6089-5p, miR-630	miR-16, miR-34a
Apigenin	miR-16, miR-138, miR-183	miR-34a, miR-423-5p, miR-103
Genistein	miR-34a, miR-145, miR-451, miR-1469, miR-132, miR-20a miR-574, miR-1296, let-7, miR-23b, miR-146a, miR-200, miR-574-3p	miR-151, miR-155, miR-221, miR-222, miR-223, miR-151, miR-1260, miR-27a, miR-21, miR-23b, miR-1260b
Gingerol	miR-27b, miR-103	
Caffeine and caffeic acid	miR-124, miR-148a	miR-9, miR-125b, miR-146a, miR-155, miR-223
Tea polyphenols	let-7b, let-7c, miR-1, miR-7-1, miR-16, miR-18, miR-25, miR-126, miR-144-3p, miR-210, miR-330, miR-34a, miR-34b, miR-92, miR-99a	miR-21, miR-30, miR-92, miR-93, miR-98-5p, miR-106b, miR-374c-5p, miR-608, miR-629, miR-449c-5p, miR-450a-2-3p, miR-453, miR-520-e
Garcinol and gemcitabine	let-7, miR-453, miR-720, miR-128, miR-200b/c, miR-638, miR-663, miR-1280	miR-21, miR-196a, miR-495, miR-494, miR-1914, miR-1977, miR-605, miR-483-3p
I3C and DIM (indole alkaloids)	let-7b/c/d/e, miR-21, miR-34, miR-146a, miR-200b/c, miR-212/132	miR-21, miR-221, miR-377, miR-31, miR-92, miR-130a, miR-146b

(continued)

Table 7.1 (continued)

	Effect on miRNA regulation	
Dietary metabolites	Upregulation of tumour suppressor miRNAs	Downregulation of oncogenic miRNAs
Mangiferin	miR-15b, miR-182	–
Rosmarinic acid	miR-124, miR-148a	miR-9, miR-125b, miR-146a, miR-155, miR-223
Ellagic acid	miR-215, miR-183, miR-375, miR-205	miR-122, miR-127, miR-214, miR-182
Indole-3 carbinol	miR-34a	miR-21
Isoflavones	miR-200b, miR-200c, let-7b, let-7e	–
Oleic acid	miR-21	–

describing the strategy in a frame of "dream or a reality" and warrant further investigations to probe the possible mechanistic involvement of xenomiRs in the physiology of recipient and cancer prevention (Motti et al. 2018).

Conclusion

Recently, miRNA research is expanding in logarithmic fashion for investing its role as diagnostic and prognostic biomarkers and even with a hope of using it as therapeutic agents.

Sizable amount of literature has accumulated in the recent past, which describes the dietary alterations modulate miRNA expression in variety of cancers. The recent preclinical data clearly suggest that recent evidence suggests that diet origin bioactive food ingredients provide protection against cancer via modulation of oncogenic and tumour suppressor miRNAs. Series of bioactive molecules such as curcumin, EGCG, sulforaphane, resveratrol, lycopene, quercetin, myricetin, luteolin, apigenin, genistein, gingerol, caffeine, caffeic acid, tea polyphenols, garcinol, gemcitabine, mangiferin, rosmarinic acid, ellagic acid, indole-3 carbinol, isoflavones and oleic acid present in variety of dietary sources are reported to upregulate the tumour suppressor miRNAs and downregulate the oncogenic miRNAs, and thereby may provide protection against cancer progression. The opportunities also lie in investigating the effect of different dietary patterns in genome-wide regulation of miRNAs in different cancers and understanding its diagnostic, prognostic and therapeutic potential.

References

Akbari KV, Khonakdar TA, Mishan MA, Tafazoli A, et al. Effects of quercetin on microRNAs: a mechanistic review. J Cell Biochem. 2019;120(8):12141–55.

Biersack B. Current state of phenolic and terpenoidal dietary factors and natural products as non-coding RNA/microRNA modulators for improved cancer therapy and prevention. Noncoding RNA Res. 2016a;1(1):12–34.

Biersack B. Non-coding RNA/microRNA-modulatory dietary factors and natural products for improved cancer therapy and prevention: alkaloids, organosulfur compounds, aliphatic carboxylic acids and water-soluble vitamins. Noncoding RNA Res. 2016b;1(1):51–63.

Chen X. MicroRNA biogenesis and function in plants. FEBS Lett. 2005;31:5923–31.

Chin AR, Fong MY, Somlo G, Wu J, Swiderski P, Wu X, Wang SE. Cross-kingdom inhibition of breast cancer growth by plant miR159. Cell Res. 2016;26:217–28.

Del Carmen Martínez-Jiménez V, Méndez-Mancilla A, Patricia Portales-Pérez D. miRNAs in nutrition, obesity, and cancer: the biology of miRNAs in metabolic disorders and its relationship with cancer development. Mol Nutr Food Res. 2018;62(1).

Devi KP, Rajavel T, Daglia M, Nabavi SF, Bishayee A, Nabavi SM. Targeting miRNAs by polyphenols: novel therapeutic strategy for cancer. Semin Cancer Biol. 2017;46:146–57. Elsevier.

Gezici S, Sekeroglu N. Regulation of microRNAs by natural products and bioactive compounds obtained from common medicinal plants: novel strategy in cancer therapy. Indian J Pharm Educ Res. 2017;51(3 Suppl):S483–8.

Momtazi AA, Shahabipour F, Khatibi S, Johnston TP, Pirro M, Sahebkar A. Curcumin as a microRNA regulator in cancer: a review. Rev Physiol Biochem Pharmacol. 2016;171:1–38.

Motti ML, D'Angelo S, Meccariello R. MicroRNAs, cancer and diet: facts and new exciting perspectives. Curr Mol Pharmacol. 2018;11(2):90–6.

Pastrello C, Tsay M, McQuaid R, Abovsky M, Pasini E, Shirdel E, Angeli M, Tokar T, Jamnik J, Kotlyar M, Jurisicova A, Kotsopoulos J, El-Sohemy A, Jurisica I. Circulating plant miRNAs can regulate human gene expression in vitro. Sci Rep. 2016;6:32773.

Philip A, Ferro VA, Tate RJ. Determination of the potential bioavailability of plant microRNAs using a simulated human digestion process. Mol Nutr Food Res. 2015;59:1962–72.

Rupaimoole R, Slack FJ. MicroRNA therapeutics: towards a new era for the management of cancer and other diseases. Nat Rev Drug Discov. 2017;16:203–22.

Samec M, Liskova A, Kubatka P, Uramova S, Zubor P, Samuel SM, et al. The role of dietary phytochemicals in the carcinogenesis via the modulation of miRNA expression. J Cancer Res Clin Oncol. 2019;145(7):1665–79.

Wang D, Fu L, Wei J, Xiong Y, DuBois RN. PPARδ mediates the effect of dietary fat in promoting colorectal cancer metastasis. Cancer Res. 2019;79(17):4480–90.

Chapter 8
Can Diet Influence Metastasis?

Abstract Cancer metastasis happens to be one of the key processes closely associated with morbidity and mortality of cancer patients. Majority of the currently used chemotherapy drugs are targeted towards primary cancer cells and the agents targeting the process of metastasis are not available and the research is evolving in this direction. A variety of dietary agents have been recently described to act as inhibitors of metastasis at various stages of its development. In the preclinical settings, variety of diet origin bioactive molecules have been explored for their role in regulation of metastasis suppressor genes, variety of metastasis signalling molecules, enzymes (especially MMPs) involved in metastasis and key metastasis-related processes such as epithelial–mesenchymal transition (EMT), invasion and angiogenesis. The preclinical and clinical literature clearly advocates that appropriate dietary modification has potential to suppress the process of metastasis in cancer patients and thereby may improve their survival and quality of life. The search of antimetastatic agents from dietary sources not only help in integrated management of cancer prevention but will also inspire the development of novel chemotherapy agents targeting metastasis.

Keywords Diet · Nutrition · Metastasis · Signalling pathways · Regulation

Abbreviations

ADAM9	A disintegrin and metalloproteases 9
AKT	Protein kinase B
Ang1	Angiotensin 1
AP-1	Activator protein 1
CCL2	Chemokine (C-C motif) ligand 2
CCN5	Cellular communication network factor 5
CD	Cluster of differentiation
COX-2	Cyclooxygenase-2
CTGF	Connective tissue growth factor
Cxcl	Chemokine (C-X-C motif) ligand

R. N. Gacche, *Dietary Research and Cancer*,
https://doi.org/10.1007/978-981-16-6050-4_8

CXCR-4	C-X-C chemokine receptor type 4 NOX: NADPH oxidase
E-cadherin	Epithelial cadherin
EGF	Epidermal growth protein
EGFR	Epidermal growth factor receptor
EMT	Epithelial–mesenchymal transition
ENPP2	Ectonucleotide pyrophosphatase/phosphodiesterase 2
Erk	Extracellular signal-regulated kinase
ET-1	Endothelin 1
FAK	Focal adhesion kinase
FGF	Fibroblast growth factor
GM-CSF	Granulocyte–macrophage colony-stimulating factor
HER2	Human epidermal growth factor receptor 2
HGF	Hepatocyte growth factor
HIF1α	Hypoxia-inducible factor 1α
HLJ1	HomoLogous to E. coli dnaJ protein 1
HO-1	Haem oxygenase 1
ICAM-1	Intercellular adhesion molecule 1
IGF	Insulin-like growth factor
Il-6	Interleukin-6
IL-8	Interleukin-8
iNOS	Nitric oxide synthase, inducible
JNK	c-Jun N-terminal kinase
KISS1	KiSS-1 metastasis suppressor
MCSCs	Metastatic cancer stem cells
MDM-2	Murine double minute 2, a p53-associated oncogene
miRNA	microRNA
MKK4	Mitogen-activated protein kinase kinase 4
MMP	Matrix metalloproteinase
MT-1	Metallothionein-1
MTA-2	Metastasis-associated protein 2
NF-kB	Nuclear factor κ B
p53	Tumour suppressor protein that protects from DNA damage
P70S6K	Ribosomal protein S6 kinase
PAI-1	Plasminogen activator inhibitor-1
PAK	p21-activating kinases
PCNA	Proliferating cell nuclear antigen
PDGF-B	Platelet-derived growth factor subunit B
PDGFR	Platelet-derived growth factor receptor
PEBP1	Phosphatidylethanolamine-binding protein 1
pFAK	Phosphorylated focal adhesion kinase
PGE-2	Prostaglandin E-2
PI3K	Phosphoinositide 3 kinase
PKC	Protein kinase C
PTEN	Phosphatase and tensin homolog

RAS	Rat sarcoma
RECK	Reversion-inducing cysteine-rich protein with Kazal motifs
RhoA	Ras homolog family member
RKIP	The Raf kinase inhibitor protein
RON	(Recepteur d'Origine Nantais) kinase
ROS	Reactive oxygen species
SNAI1	Snail family zinc finger 1
SPARC	Secreted protein acidic and rich in cysteine
Src	Rous sarcoma oncogene cellular homolog
STAT-3	Signal transducer and activator of transcription 3
TGF	Transforming growth factor
TGF-α	Transforming growth factor-α
TGF-β	Transforming growth factor-β
TIMP	Tissue inhibitor of metalloproteinase
TIMP1	TIMP metallopeptidase inhibitor 1
TNF-α	Tumour necrosis factor-α
uPA	Urokinase
uPAR	Urokinase plasminogen activator surface receptor
VCAM-1	Vascular cell adhesion molecule 1
VEGF	Vascular endothelial factor
VEGFR	VEGF receptor
VWF	von Willebrand factor
ZEB1	Zinc finger E-box-binding homeobox 1

Introduction

Can Diet Influence Metastasis?

Metastasis is a process of spread and colonization of cancer cells through blood and/or lymph vessels to distant tissues and organs in the body, and it has been considered as major factor in the morbidity and mortality of cancer patients. Since long time, it was believed that tumours need to attain a particular size that further penetrates into localized vessels before they release cells into the circulation. However, the advances in the molecular techniques proved this belief as an illusion and now we have a renewed understanding that cancerous tumours may shed their cells into the blood or lymph circulation from the earliest stages of growth, which may not be size-dependent. Thanks to the impressive repertoires of immune surveillance, which kill majority of the cells during circulation or before they land at particular niche and colonize (Friberg and Nyström 2015). Besides the great strides of efforts of past two decades and advent of effective immunotherapies, metastasis has remained a poor patient prognosis and a prominent cause of over 90% cancer-related mortality. The anticancer drug discovery and developmental programmes including the molecularly tailored targets and recent immunotherapies generally ignore or do not include the molecular targets of metastasis inhibition. As a result, majority of the currently prescribed chemotherapy drugs are targeted against the primary tumour cells and there is lack of effective anticancer agents, which selectively target metastasis (Lambert et al. 2017). Of note, metastatic tumours demonstrate a remarkable heterogeneity and different biology than that of its primary tumour counterpart. Therefore, upcoming cancer drug developing endeavours should capitalize more on inhibiting the fundamental metastatic processes that principally do not rely on primary tumour responses (Anderson et al. 2019).

Although several molecular underpinnings of metastasis are being explored, still the pathophysiological characteristics and druggable molecular signatures that globally arrest the metastasis are yet to be evolved. Per say, the metastatic process appears to be complex, multistep and guided cascade driven by the cancer cells. The metastatic signatures start with disruption or breaking of the extracellular matrix basement membrane and dissociating from the parent tumour body followed by entering into the blood or lymph circulation, sustaining in the blood or lymph circulation, reaching at the capillary front of the target tissue or organ and finally landing at the distant site and attaining the progressive colonial growth and survival abilities. The most significant two intrinsic properties required by a cancer cell to undergo successful metastasis are the invasiveness and anchorage-independent survival. Both these processes come under the control of gene regulation (Steeg 2016).

Targets for Arresting Metastasis

The past couple of decade's researches have contributed significantly in making the framework of targeting metastasis. Among the major factors and pathways that recruits metastasis and which are potential drug targets include the integrins, selectins, immunoglobulin superfamily receptors, chemokine receptors, carbohydrate-rich proteins, matrix metalloproteinases (MMPs), proteinases of the fibrinolytic system, cathepsins, elastase, heparanase, homing receptors, peptide motifs in the vasculature, addressins and mobilization of bone marrow cells. These molecular players are mostly proactive at the initial stages of metastasis helping the cells for homing and invasion. In brief, looking towards the involvement of multiple factors and complex orchestration of metastasis, the most important pro-metastasis players such as integrins, CD44 (a glycoprotein conferring metastasis), chemokine receptors, TGF-α, MMPs, bone marrow-derived cells supporting metastasis and angiogenic factors (mostly VEGF, VEGF receptors and FGF-2) have been proposed as prospective antimetastatic drug targets and developing selective inhibitors against these key proteins might lead to development of antimetastasis therapeutic regime (Weber 2017; Anderson et al. 2019). Although few inhibitors against the above-mentioned proteins are available, but they are not sufficient to combat the complexity of progressive metastasis and to extend the apparent clinical benefits to patients with advanced cancer. The preclinical and clinical studies on molecular mechanisms of metastasis are evolving (Sleeman 2018), but at the same time there is need of employing advanced molecular approaches for searching and developing effective candidate drugs to counteract the novel targets of metastasis (Anderson et al. 2019).

One more aspect that has invited the curiosity of scientific community is the role of "epithelial–mesenchymal transition" (EMT) in pathophysiology of cancer in general and metastasis is particular. In fact, classically EMT is a reversible switch implicated in phenotypic transformation from epithelial to mesenchymal cells, which is a compatible phenotype for invasion and migration. The recent findings have extended the EMT process to a more comprehensive functional repertoire in tumour progression and augmented it as one of the basic prerequisites of cancer invasion and metastasis. The current evidence indicates that EMT also plays a crucial role in metabolic reprogramming, stemness, immune evasion and drug resistance of cancer cells. The molecular underpinning of EMT in relation to tumorigenesis, metastasis and several other aspects of cancer progression is consistently investigated with a hope of developing EMT as an attractive target for cancer therapeutics. Series of factors and signalling pathways like TGF-β, EGFR, TβRI, ALK1, HGF/c-Met, Frizzled, Notch and Wnt are described as drivers of EMT and possibly the targets for arresting EMT or reverting EMT into MET: mesenchymal–epithelial transition (Fischer et al. 2015; Cho et al. 2019).

Can diet influence metastasis and cancer stem cells? is an important question that needs to be considered while addressing the impact of diet on various aspects of cancer progression and tumorigenesis. Before we discuss the counteracting activities of dietary substances as inhibitors of pro-metastasis factors, signalling pathways, EMT and MCSCs, it is imperative to take a cursory look, as to what is the role of specific diet or dietary patterns in metastasis. The preclinical research data of past and present findings link the increased risk of metastasis with high-fat diet (HFD) in variety of cancers (Erickson and Hubbard 1990; Chambers 2009; Sundaram and Yan 2016). The molecular aspects of high-fat-driven induction of metastasis also support the positive influence of HFD in accelerating the metastasis. For example, HFD feed mice with squamous cell carcinoma developed more lymph node metastases as compared to mice feed with normal diet. The increased metastatic ability was found to be associated with expression of the fatty acid receptor CD36 in metastasis-initiating cancer cells (Pascual et al. 2017). In another animal model study, HFD-induced metastasis rate was found to be related to loss of PTEN and PML genes in prostate cancer (Chen et al. 2018).

An interesting animal model studies have demonstrated that in contrast to omega-6 fatty acids, isocaloric and isolipic long-chain omega-3 fatty acid consumption was observed to suppress the tumour growth and metastasis formation in the 4T1 animal model studies of metastatic breast cancer (Khadge et al. 2018). In a recent preclinical animal model study, a ligand-dependent transcription factor PPAR-δ involved in fatty acid metabolism mediates the effect of dietary fat in accelerating metastasis in colorectal cancer (Wang et al. 2019). Supportive to these findings, obesity has been reported to alter the landscape of lung myeloid cell and increase breast cancer metastasis via IL5 and GM-CSF (Quail et al. 2017). Also, caloric restriction and ketogenic diet have demonstrated antimetastatic effects in animal model experiments (Lv et al. 2014). The recent studies also demonstrated that dietary restriction or reduced bioavailability of single amino acid like asparagine reduces metastasis without altering the growth of the primary tumour. Opposite to this, increased inclusion of dietary asparagine or upregulated expression of asparagine synthetase promotes metastasis in several preclinical settings. In fact, such studies are significant in demonstrating how the dietary restriction or bioavailability of a single amino acid is instrumental in regulating metastatic progression (Knott et al. 2018).

Now, in the framework of analysing the dietary metabolites as inhibitors of pro-metastasis factors, signalling pathways, EMT and MCSCs, a cursory look at the current literature shows that variety of dietary metabolites have potential to interact with antimetastasis drug targets and also help to suppress the cancer metastasis genes (Meadows 2012; Weng and Yen 2012a, b; Singh et al. 2014; Yen et al. 2018). Variety of dietary phenolic acids, monophenol, polyphenols such as gallic acid (red fruits, black radish, onions and tea), chlorogenic acids (coffee, apples, pears, eggplant, blueberries, tomatoes, strawberries, potatoes, etc., most commonly supplemented in the form of green coffee extract), carnosol (rosemary), caffeic acid (coffee, apples, artichoke, berries and pears), capsaicin (alapeño peppers, cayenne peppers and other chilli peppers), 6-gingerol and 6-shogaol (ginger) are extensively reviewed for their anti-invasive and antimetastatic effects in in vitro and in vivo preclinical model studies. These dietary substances act as antimetastatic agents by modulating the activities of several factors and metastasis-related pathways like PKC, FAK and Cdc42, and by phosphorylation of Ras, ERK, JNK, Rac1, p38, PI3K, PAK1, Akt, p70S6K, RhoA and RhoB signalling, which in turn downregulate the expression of AP-1, STAT3 and NF-kB. These dietary molecules have also demonstrated antimetastatic activity by downregulation of MMP-1–3, MMP-9, and MMP-14, MT-1, uPA/uPAR, actin stress fibre formation, F-actin, integrin, NCAM, CXCR4, CCL2, IL-6 and IL-8, ENPP2, ICAM-1, RON, b-FGF, COX-2, HER2, ROS, SPARC, VCAM-1, HGF, VEGF/EGF/EGFR, HO-1, HIF-1a and ADAM17 and also by upregulation of HLJ1, E-cadherin, TIMP-1, PAI-1, maspin, b-catenin, filopodia and tensin in different tumour cells or animal model studies (reviewed by Weng and Yen 2012b).

Dietary Metabolites as Potential Antimetastatic Agents

Perhaps the most widely investigated dietary bioactive class having multiple biological activities is the flavonoids, which comprises of subclasses like flavones, flavonols, flavanones, flavanonols, flavanols, isoflavones and anthocyanidins. The most widely investigated and in-depth studied dietary flavonoids possessing impressive antimetastatic potential are green tea flavonoids like catechin, epigallocatechin, epigallocatechin-3-gallate, epicatechin-3-gallate and epicatechin. While describing the mechanism of action, these green tea substances suppress the expression of IGF-1, ET-1, MT1-MMP and uPA leading to downregulation of MMP-2, MMP-9 and other MMPs involved in metastasis. The antimetastatic activities are also associated with inhibition of angiogenesis by inhibiting VEGF-A, VEGF, VEGFR, IL-8, HIF-1α along with downregulation of β-1 integrin signalling pathway, laminin receptor (67LR), CD44 and CD54, which are key players in cellular adhesion during metastasis process. Green tea flavonoids also target the EMT and invasion players by overexpressing the E-cadherin, cytokeratin and desmoglein 2 and downregulating the mesenchymal biomarkers like N-cadherin, vimentin, fibronectin, snail, pFAK, pSrc, HGF-Met signalling, RKIP and CCN5 (Khan and Mukhtar 2010; Bag and Bag 2020). The other flavonoids targeting different metastasis targets include the genistein and genistin, silibinin, quercetin, anthocyanin, myricetin, tangeretin, luteolin, apigenin, kaempferol, glycitein, daidzein and naringenin. The flavonoids are reported to manifest their anti-invasive and antimetastatic abilities by upregulation of E-cadherin, MTA3, PAI-1, γ-catenin, TIMP-1/TIMP-2, RECK, KAI1, PNII, alpha 1-AT, β1-integrin, cytokeratin-18 and OPG, and also by suppressing the expression levels of MMPs (2, 3, 7, 9, 12), MT1-3-MMP, uPA, uPAR, tPA, MUC1, vimentin, vitronectin, fibronectin, snail, EGF, EGFR, VASP, ErbB2/ErbB3, Met (HGFR), EMMPRIN, VEGF, VEGR-R2, HIF-1α, β1-/β4-integrin, β-catenin, PSA, angiopoietin1/2, α5-/β1-/αv-/β3-integrin receptors, CXCR4, CXCL12, OPN, MDM-2, COX-2, PGE2, iNOS, vWF, PECAM-1 (CD31), claudin, plasmin activation, RANKL and osteoclast (reviewed by Weng and Yen 2012a).

Apart from dietary phenolic acids, polyphenols and flavonoids, there are several other dietary nonphenolic bioactive metabolites showing impressive antimetastatic effects. The dietary metabolites belonging to alkaloids like caffeine (coffee and cocoa beans, kola nut, tea), piperine (black pepper); indol compounds such as indole-3-carbinol (cabbage, cauliflower), 3,3′-diindolylmethane (broccoli sprouts, kale); organosulphur substances like allicin, *s*-allylcysteine, diallyl sulphide, diallyl disulphide (garlic), sulforaphane (mustard green, cabbage, broccoli), benzyl isothiocyanate (Brussels sprouts), phenethyl isothiocyanate (broccoli) and allyl isothiocyanate (horseradish); carotenoids such as beta-carotene (carrot), lutein (corn, spinach, orange) and lycopene (tomato, watermelon); variety of dietary terpenoids like retinoic acid (carrot), ursolic acid (cranberry), limonene (lime, lemon, tangerine) and zerumbone (Asia ginger); and furocoumarins molecules like bergamottin (grapefruit juice) and lignin substance such as sesamin (sesame) have demonstrated anti-invasion and antimetastasis activities in series of in vitro and in vivo model studies. The nonphenolic bioactive metabolites mediate their anti-invasive and antimetastasis capabilities by upregulating variety of metastasis-suppressing genes such as E-cadherin, nm23-H1, TIMP-1,2,3, PAI-1, PTEN, ROS, VEGF, miR-194, miR-146a, miR-34a, Notch1 and Notch2, IL-2, IL-6, IL-8 and TGF-b and also by suppressing the expression of pro-metastasis proteins such as vimentin, MMP-1–MMP-3, MMP-7, MMP-9, MMP-13, MMP-14, uPA, uPAR, MT1-MMP, a5, av, b1, b3-integrins, EGFR, PDGF, VEGF, VEGFR2, HGF-c-met, IL-1b,4,6,8, ROS, NOX4, NO, iNOS/NOS2, CXCR4, CXCL12, TGF-b1, ICAM-1, COX-2, snail, TNF-a, IFN-g, claudin-2–4, miR-200c, miR-92a, TIMP-1, VCAM-1, MTA-2, galectin-3, PCNA, CD34, CD44, CD44v6, CD31, Ang-1, Ang-2 and Tie-2, and by modulation of p53 (reviewed by Yen et al. 2018). The outline of inhibition of metastasis-related processes by dietary interventions is shown in Fig. 8.1.

More precisely, series of preclinical animal model studies have described the role of diet, nutritional and dietary constituents in regulating the metastasis suppressor genes. For example, luteolin upregulated the expression of E-cadherin protein in PC3 prostate cancer cells and these expression levels were found to be associated with decreased cell invasion. Luteolin also upregulated protein expression in PC3 tumour extracts injected onto the dorsal flank of in BALB/cA-nu (nu/nu) mice, and it also inhibited lung metastasis (Zhou et al. 2009). Lycopene, a dietary bioactive molecule from tomato, increased mRNA levels and upregulated the expression of the NM23-H1 protein in highly invasive SK-Hep-1 hepatic adenocarcinoma cells. The findings of the experiments correlated with the inhibition of migration and invasion (Huang et al. 2005). Fish oil consumption was found to be associated with downregulation of mRNA and expression of CD44 protein in MDA-MB231 breast tumours after intracardiac inoculation into mice. The findings were correlated with the reduced bone metastasis process (Mandal et al. 2010).

Pomegranate juice induced the expression of E-cadherin protein in DU145 prostate cancer cells. It also promoted the adhesion to gelatine-coated plates and arrested the migration and chemotaxis of DU145 and PC3 prostate cancer cells (Wang et al. 2011). A dietary component curcumin significantly decreased B16-F10 lung metastasis by eightfold in C57BL/6 mice (Ray et al. 2003). Series of dietary flavonoids such as 7-hydroxyflavone, baicalein and 4′,-5,7-trihydroxy-flavone downregulated the expression of TIMP-2 protein SCC-4 human tongue squamous cell carcinoma cells. All these flavonoid bioactive molecules inhibited migration, invasion and intravasation in the chicken chorioallantoic membrane model studies (Yang et al. 2008). Variety of dietary bioactive molecules belonging to polyphenols, proteins, amino acids, flavonoids, vitamins, glucosinolates and carotenoids (lycopene) demonstrated interaction with series of metastasis-related tumour suppressor genes such as E-cadherin, NM23, TIMP-1, CD44, KISS1, KAI1, MKK4, PEBP1, RKIP, RECK, CTGF and RhoGDI2 (reviewed by Meadows 2012).

As an important framework of inhibiting metastasis, the quest of finding dietary inhibitors of EMT is one of the thrust areas for developing dietary interventions for cancer prevention. Series of dietary substances such as quercetin, luteolin, tangeretin, 3,6-dihydroxyflavone, curcumin, emodin, fisetin, procyanidin C1, silibinin, chrysin and tea polyphenols like catechin, epicatechin, epigallocatechin, epicatechin-3-gallate and epigallocatechin-3-gallate are described as EMT inhibitors by inhibiting several EMT-related proteins (Weng and Yen 2012a; Kim et al. 2016). Among variety of phytochemicals tested as possible inhibitors of EMT, phytoestrogens, which are mostly found in soya foods, vegetables and fruits, are considered as significant sources of EMT inhibitors. Phytoestrogens generally have distinct cancer-protective properties against oestrogen-related cancers. The most extensively studied dietary phytoestrogens as EMT inhibitors are genistein, kaempferol, resveratrol and 3,3′-diindolylmethane (DIM). Mechanistically, the anti-EMT effects of phytoestrogens, flavonoids and polyphenols are manifested by interactions with variety of EMT targets and perturb the EMT process by

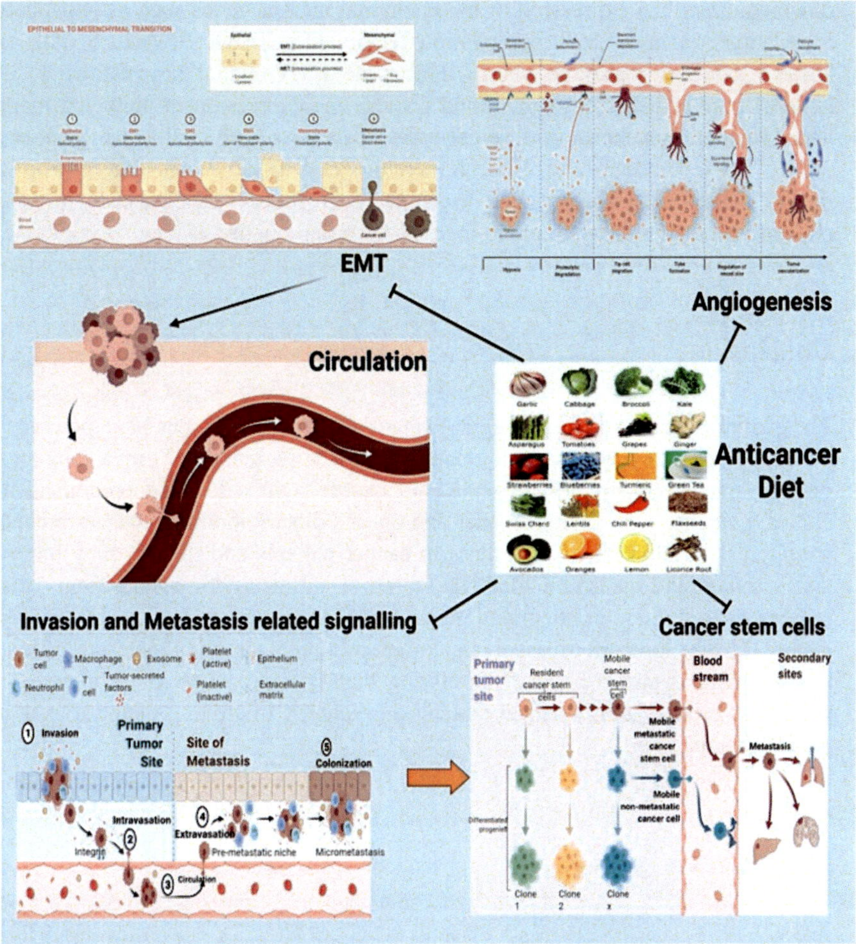

Fig. 8.1 Process of metastasis initiates with local invasion wherein the cancer cell connected ECM is digested by MMPs. Further, the cancer cells start migration; in this step, one of the important steps is the EMT that permits the cancer cells to achieve migration and invasion. In fact, the EMT is induced by various molecular cues such as hypoxia, cytokines and growth factors and most importantly by transcriptional factors such as Twist, Snail, Slug, ZEB1 and ZEB2 or by epigenetic regulation. EMT phenotype is associated with downregulation of E-cadherin and upregulation of N-cadherin and vimentin. Tumour cells escape from immune response in a very inscrutable manner and establish a tumour-supportive microenvironment (premetastatic niche) at the site of future metastasis. Metastatic process is strongly associated with the tumour neovascularization or angiogenesis. The process of angiogenesis is mostly triggered by secretion of pro-angiogenic factors such as VEGF, FGF-2, ILs, PlGF, TGF-β, PDGF, HGF and angiopoietins. The dietary bioactive molecules are known to attenuate the process of EMT by downregulating the marker proteins/transcription factors needed for successful EMT, and these molecules effectively inhibit the process of angiogenesis at various stages. Interestingly, the dietary bioactive molecules and also described to arrest the invasion by inhibiting the MMPs and are also described to reduce the population of cancer stem cells

downregulating the expression of mesenchymal marker genes such as N-cadherin, Slug, vimentin and EMT transcription factors like ZEB1, SNAI1, SNAI2 and TWIST. They are also known to inhibit the EMT-related key pathways, like Notch-1 and TGF-beta signalling and activate the expression of epithelial marker genes such as E-cadherin in different types of cancers (Lee et al. 2016; Kim et al. 2016). The preclinical and clinical literature clearly advocates that appropriate dietary modification has potential to suppress the process of metastasis in cancer patients and thereby may improve their survival and quality of life.

Conclusion

The evolving literature clearly suggests that the dietary molecules have potential to positively regulate the metastasis-related processes such as EMT, invasion, angiogenesis, metastasis suppressor genes and cancer stem cells. The preclinical and clinical evidence also advocates that appropriate dietary modification has potential to suppress the process of metastasis in cancer patients and thereby may improve their survival and quality of life. The search of antimetastatic agents from dietary sources not only help in integrated management of cancer prevention, but will also inspire the development of novel chemotherapy agents targeting metastasis. Nevertheless, there is also a hope of modifying dietary approaches for suppression of metastasis process and in cancer patients and thereby improving their survival and quality of life.

References

Anderson RL, Balasas T, Callaghan J, Coombes RC, Evans J, Hall JA, et al. A framework for the development of effective anti-metastatic agents. Nat Rev Clin Oncol. 2019;16(3):185–204.

Bag N, Bag A. Antimetastatic properties of tea polyphenols. Nutr Cancer. 2020;72(3):365–76. https://doi.org/10.1080/01635581.2019.1638426.

Chambers AF. Influence of diet on metastasis and tumor dormancy. Clin Exp Metastasis. 2009;26: 61–6.

Chen M, Zhang J, Sampieri K, Clohessy JG, Mendez L, Gonzalez-Billalabeitia E, Liu XS, et al. An aberrant SREBP-dependent lipogenic program promotes metastatic prostate cancer. Nat Genet. 2018;50(2):206–18.

Cho ES, Kang HE, Kim NH, Yook JI. Therapeutic implications of cancer epithelial-mesenchymal transition (EMT). Arch Pharm Res. 2019;42:14–24.

Erickson KL, Hubbard NE. Dietary fat and tumor metastasis. Nutr Rev. 1990;48(1):6–14.

Fischer KR, et al. Epithelial- to-mesenchymal transition is not required for lung metastasis but contributes to chemoresistance. Nature. 2015;527:472–6.

Friberg S, Nyström A. Cancer metastases: early dissemination and late recurrences. Cancer Growth Metastasis. 2015;8:43–9.

Huang CS, Shih MK, Chuang CH, et al. Lycopene inhibits cell migration and invasion and upregulates Nm23-H1 in a highly invasive hepatocarcinoma, SK-Hep-1 cells. J Nutr. 2005;135(9):2119–23.

Khadge S, Thiele GM, Sharp JG, et al. Long-chain omega-3 polyunsaturated fatty acids decrease mammary tumor growth, multiorgan metastasis and enhance survival. Clin Exp Metastasis. 2018;35(8):797–818.

Khan N, Mukhtar H. Cancer and metastasis: prevention and treatment by green tea. Cancer Metastasis Rev. 2010;29:435–45.

Kim EK, Choi EJ, Debnath T. Role of phytochemicals in the inhibition of epithelial–mesenchymal transition in cancer metastasis. Food Funct. 2016;7(9):3677–85.

Knott SRV, et al. Asparagine bioavailability governs metastasis in a model of breast cancer. Nature. 2018;554:378–81.

Lambert AW, Pattabiraman DR, Weinberg RA. Emerging biological principles of metastasis. Cell. 2017;168(4):670–91.

Lee GA, Hwang KA, Choi KC. Roles of dietary phytoestrogens on the regulation of epithelial-mesenchymal transition in diverse cancer metastasis. Toxins. 2016;8(162):1–17.

Lv M, Zhu X, Wang H, Wang F, Guan W. Roles of caloric restriction, ketogenic diet and intermittent fasting during initiation, progression and metastasis of cancer in animal models: a systematic review and meta-analysis. PLoS One. 2014;9:e115147.

Mandal CC, Ghosh-Choudhury T, Yoneda T, et al. Fish oil prevents breast cancer cell metastasis to bone. Biochem Biophys Res Commun. 2010;402(4):602–7.

Meadows GG. Diet, nutrients, phytochemicals, and cancer metastasis suppressor genes. Cancer Metastasis Rev. 2012;31(3–4):441–54.

Pascual G, et al. Targeting metastasis-initiating cells through the fatty acid receptor CD36. Nature. 2017;541:41–5.

Quail DF, et al. Obesity alters the lung myeloid cell landscape to enhance breast cancer metastasis through IL5 and GM-CSF. Nat Cell Biol. 2017;19:974–87.

Ray S, Chattopadhyay N, Mitra A, et al. Curcumin exhibits antimetastatic properties by modulating integrin receptors, collagenase activity, and expression of Nm23 and E-cadherin. J Environ Pathol Toxicol Oncol. 2003;22(1):49–58.

Singh BN, Singh HB, Singh A, Naqvi AH, Singh BR. Dietary phytochemicals alter epigenetic events and signalling pathways for inhibition of metastasis cascade: phytoblockers of metastasis cascade. Cancer Metastasis Rev. 2014;33(1):41–85.

Sleeman JP. Dietary regulation of metastasis. Clin Exp Metastasis. 2018;35:713–4.

Steeg PS. Targeting metastasis. Nat Rev Cancer. 2016;16:201–18.

Sundaram S, Yan L. High-fat diet enhances mammary tumorigenesis and pulmonary metastasis and alters inflammatory and angiogenic profiles in MMTV-PyMT mice. Anticancer Res. 2016;36 (12):6279–87.

Wang L, Alcon A, Yuan H, et al. Cellular and molecular mechanisms of pomegranate juice-induced anti-metastatic effect on prostate cancer cells. Integr Biol (Camb). 2011;3(7):742–54.

Wang D, Fu L, Wei J, Xiong Y, DuBois RN. PPARδ mediates the effect of dietary fat in promoting colorectal cancer metastasis. Cancer Res. 2019;79(17):4480–90.

Weber GF. Molecular genetics of metastasis. In: eLS. Chichester: Wiley; 2017. https://doi.org/10.1002/9780470015902.a0022447.pub2.

Weng CJ, Yen GC. Flavonoids, a ubiquitous dietary phenolic subclass, exert extensive in vitro anti-invasive and in vivo anti-metastatic activities. Cancer Treat Rev. 2012a;31:323–51.

Weng CJ, Yen GC. Chemopreventive effects of dietary phytochemicals against cancer invasion and metastasis: phenolic acids, monophenol, polyphenol, and their derivatives. Cancer Treat Rev. 2012b;38:76–87.

Yang SF, Yang WE, Kuo WH, et al. Antimetastatic potentials of flavones on oral cancer cell via an inhibition of matrix-degrading proteases. Arch Oral Biol. 2008;53(3):287–94.

Yen GC, Tsai CM, Lu CC, Weng CJ. Recent progress in natural dietary non-phenolic bioactives on cancers metastasis. J Food Drug Anal. 2018;26:940–64.

Zhou Q, Yan B, Hu X, et al. Luteolin inhibits invasion of prostate cancer PC3 cells through E-cadherin. Mol Cancer Ther. 2009;8(6):1684–91.

Chapter 9
Impact of Dietary Interventions on Cancer Stem Cells

Abstract The researches in past one decade have identified cancer stem cells (CSCs) as a major culprit of recurrence, heterogeneity, metastasis, drug and radioresistance in cancer treatment. With the increasing awareness of the significance of a dietary intervention in human health, the research is also coevolving with the molecular understanding of association between diet and cancer. Cancer stem cells are a subpopulation of cancer cells with unique characteristics of self-renewal potential, aggressive invasiveness, promoting metastasis and demonstrated resistance against chemo- and radiotherapy. The preclinical data have clearly demonstrated the anticancer effects of dietary substances against CSCs. Series of dietary ingredients inhibit the CSC-associated signalling pathways and ultimately adversely affect maintenance of CSC stemness, survival and proliferation. To integrate the anti-CSCs dietary molecules with chemotherapy drugs, there is need to establish the safety and efficacy of these molecules by performing relevant clinical trials.

Keywords Diet · Nutrition · Cancer · Stem cells · Recurrence · Relapse · Drug resistance

Abbreviations

AKT	Protein kinase B
CAR-T cells	Chimeric antigen receptor T cells
Caspase	Cysteine proteases with aspartate specificity
CD	Cluster of differentiation
CSC	Cancer stem cell
EGCG	Epigallocatechin-3-gallate
HDAC	Histone deacetylase
JAK/STAT	Janus kinase/signal transducer and activator of transcription
LSC	Leukaemic stem cell
MCSCs	Metastatic cancer stem cells
MDR1	Multidrug-resistant protein 1
miRNA	microRNA

R. N. Gacche, *Dietary Research and Cancer*,
https://doi.org/10.1007/978-981-16-6050-4_9

MRP1	Multidrug resistance-associated protein 1
MT-1	Metallothionein-1
MTA1	Metastasis-associated 1
mTOR	Mammalian target of rapamycin
NF-kB	Nuclear factor κ B
PARP	Poly (ADP-ribose) polymerase
PI3K	Phosphoinositide 3 kinase
PPAR	Peroxisome proliferator-activated receptor
PTEN	Phosphatase and tensin homolog
RAS	Rat sarcoma
STAT-3	Signal transducer and activator of transcription-3
TGF	Transforming growth factor

Introduction

Cancer stem cells (CSCs) constitute a small subpopulation or a fraction of cells residing within tumours and possess potential for differentiation, self-renewal and tumorigenicity. CSCs are mostly identified and enriched based on the presence of several cell surface markers like CD24, CD44, CD133 and nestin. A regulatory framework consists of certain factors and signalling pathways viz. Notch, Wnt/β-catenin, Hedgehog signalling, JAK-STAT, PI3K/AKT/mTOR, TGF, SMAD and PPAR, which basically control the functional properties of CSC. Cancer stem cells develop resistance to the conventional chemotherapy drugs, radiation treatment, heterogeneity, metastasis and recurrence of cancer condition. Taking into the consideration, it is necessary to develop an innovative anticancer therapeutic agents and great strides of efforts are in progress in this direction. In fact, some of the anti-CSC small molecules, antibodies, vaccines, and CAR-T cells have been tailored against CSCs, and some of these therapeutic agents are being studied at various phases of clinical trials (Yang et al. 2020).

The strong potential of self-renewal is a characteristic of CSCs where it either follows symmetrical cell division to generate two CSCs or form one CSC and one daughter cell, which further attributes for the tumorigenesis. The CSCs follow the symmetrical cell division either generates two CSCs or forms one CSC and one daughter cell. By adopting symmetrical cell division, CSCs produce excessive cell growth and ultimately contribute in tumour formation (Hill 2006). Searching the dietary substances regulating the self-renewal ability of CSCs might be a key issue in arresting the CSC-mediated tumorigenesis. Another opportunity is to identify diet origin bioactive molecules, which specifically inhibit the CSC self-renewal regulating pathways, which otherwise are dysregulated and promote the uncontrolled division of CSCs, which ultimately lead to tumorigenesis (Jin et al. 2017). Epithelial-to-mesenchymal transformation (EMT) has been described as a basic prerequisite for metastasis, especially the invasion of tumour cells to distant places (Le et al. 2008).

Nevertheless, EMT is also reported to generate a population of cells possessing stem cell-like properties (Mani et al. 2008). Recent data show that the CSCs are closely associated with EMT, and EMT happens to be a key process for tumour invasion and metastasis (Yang et al. 2020).

Furthermore, current anticancer research is also curious about understanding the role and mechanism of CSC-mediated drug resistance as it happens to be instrumental for effective cancer treatment and circumventing recurrence (Garcia-Mayea et al. 2020). Interestingly, CSCs efficiently express the variety of multidrug resistance (MDR) proteins such as ATP-binding cassette transporters including MDR1, MRP1 and ABCG2. These MDR proteins protect leukaemia cells, and other cells from solid tumour from drug-induced damage, which further leads to the development of drug resistance (Gottesman et al. 2002). Moreover, the current research has identified that CSCs acts as a key "seeds" for tumour initiation, development, metastasis and its recurrence (Cojoc et al. 2015). The evolving tumour heterogeneity is of the emerging limitation for chemotherapy drugs and the population of CSCs is described to be highly heterogeneous in nature and therefore difficult to treat (Tang 2012). Therefore, it will be interesting to search the dietary substances/sources, which may circumvent the CSC-mediated chemoresistance and evolving tumour heterogeneity.

Impact of Dietary Interventions on Cancer Stem Cells

The advances in CSC research have clearly linked its role in metastasis (metastatic cancer stem cells: MCSCs) and recurrence of cancer conditions after drug holidays. The studies on hierarchical tumour model have suggested that the CSCs are associated with the origin of tumour formation, metastasis and relapse of cancer. The advancement in the scientific study in the past two decades has established a conclusive evidence about the presence of CSCs. Owing to consistent accumulating evidence towards the existence of CSCs in variety of solid tumours of cancers (Visvader and Lindeman 2008), the CSC hypothesis is inspiring the scientific community and turned out to be a fundamental underpinning of cancer biology research and therapy. The accumulated CSC-mediated molecular mechanisms favouring the cancer growth compel to the researchers to tailor the treatment against CSCs for the effective management of cancer.

The recent investigations have identified a subpopulation of MCSCs possessing stemness and invasive abilities and also promote cancer metastasis. MCSCs are derived from the population of invasive cancer stem cells, which are capable to proliferate and develop into cancer tissues. MCSCs are distinct from stationary cancer stem cells in many functional aspects; for example, they can invade normal vascular tissues, demonstrated resistance to chemotherapy and radiotherapies. Interestingly, MCSCs can escape from immune surveillance and survive in circulation and promote metastasis (Liao et al. 2014). In fact, MCSCs reside within a specific niche and has the crosstalk with surrounding microenvironment to regulate homeostasis and morphogenesis. The MCSC niche is characterized by hypoxia, oxidative stress, low pH and inflammation, which are responsible for maintaining stemness of MCSCs and targeting MCSC niche may also help to counteract the functions of MCSCs (Liao et al. 2014; Mihaylova et al. 2014). The inscrutable pro-metastatic potentials of MCSCs make it as potential therapeutic targets for cancer metastasis. On the eve of emerging roles of MCSCs in metastasis, drug resistance, self-renewal and recurrence after drug holidays, the future settings of cancer drug development are also focused towards targeting cancer stem cells (CSCs) and searching effective dietary anti-CSCs or MCSCs agents is a part of making dietary interventions as complementary and alternative cancer therapeutic approach.

CSCs are found to be instrumental in initial cancer formation, growth, proliferation and also recurrence of cancer after chemotherapy. It has been well-established fact that CSCs demonstrate more resistance towards conventional therapeutic approaches as compared to differentiated cancer cells (Creighton et al. 2009). Therefore, searching the bioactive molecules from the dietary sources, which will target the anticancer drug targets and may specifically target CSCs and avoiding toxicity to normal stem cells, is appreciated as alternative and complementary approach for the management of cancer. Considering the unique features (Saito et al. 2010; Lathia et al. 2011) and dynamics (Fornari et al. 2014) of CSCs, it can be targeted by several approaches such as inhibition of self-renewal, initiation of differentiation into mature cancer cells, and complementing the anticancer agents.

The recent research describes that the nutritional microenvironment is one of the important factors that contribute towards the role of CSCs in homeostasis and tumorigenesis (Li et al. 2019). In the same line of research, a high-fat diet microenvironment harbours a greater number of stem cells, enhances stemness and also promotes tumorigenesis via PPAR-δ activation in intestinal stem cells (Beyaz et al. 2016). A recent study has demonstrated that higher levels of amino acids and its metabolism are a prerequisite in the survival of de novo leukaemia stem cells (LSCs). The metabolism of amino acid drives the oxidative phosphorylation, which is crucial for survival of LSCs. Interestingly, the LSCs obtained from relapsed human acute myeloid leukaemia patients were independent of the amino acid metabolism and owed to increased fatty acid metabolism. The outcome of this study elucidates that contrasting to healthy or mature cancer cells, the CSCs sustain cancer are dependent on energy from protein or from amino acids. This unique dietary preference makes CSCs susceptible to drugs that interfere with amino acid uptake. Perhaps, strategically such types of drugs may be toxic to CSCs leaving healthy cells unaffected (Jones et al. 2018). Of note, a combination of dietary phytochemicals (curcumin, resveratrol, EGCG, genistein) and repositioned drugs (metformin, niclosamide, thioridazine, chloroquine) has been described as a novel therapeutic to increase toxicity against CSCs (Chan et al. 2018).

Series of dietary bioactive molecules have been reported to inhibit the CSC-related signalling pathways; for instance, EGCG, a polyphenol of tea, inhibits

Wnt/β-catenin signalling and implicates the self-renewal and invasive abilities of certain set of CSCs (Mineva et al. 2013; Lee et al. 2013). Several dietary phytochemicals (resveratrol, curcumin, genistein, diallyl trisulphide, pterostilbene, sulforaphane, epigallocatechin-3-gallate, phenethyl isothiocyanate), plant-based functional foods (*Syzygium aromaticum*, Chinese bayberry leaf, green algae capsosiphon fulvescens, *origanum vulgare*), herbal extracts (*Trianthema portulacastrum*, *Punica granatum*, *Gynura divaricata*, *Spatoglossum asperum*, *Padina tetrastromatica*, *Hormophysa triquetra*), and a cocktail of resveratrol and grape seed extracts demonstrated the anticancer effects. In preclinical studies, their anticancer effects are mainly because of its ability to target various CSC-related markers (CD133, CD44, CD24, CD13, CD90, etc.) and pathways (Wnt/beta-catenin, Notch, Hedgehog, PI3K/Akt, JAS/STAT) (Liskova et al. 2019).

A large number of plant-based bioactive molecules obtained from vegetables, fruits and grains have ability to inhibit of CSCs hence showed promising anticancer properties (Nobili et al. 2009; Graham et al. 2000). However, series of phytochemicals have been described to interact adversely with signalling pathways (mostly hedgehog, Wnt/β-catenin, and Notch-) that are associated stemness maintenance of CSCs and regulate the CSC phenotype, self-renewal and differentiation (Kim and Surh 2015; Dandawate et al. 2013; Kim et al. 2012). Retinoic acid, an active molecule derived from vitamin A in animals, has been reported to differentiate CSCs or hamper its formation in glioblastoma by inhibiting the Notch signalling pathway. Vitamin D or its analogue derivative is known to downregulate its Wnt/β-catenin and Notch signalling and thus induce CSC differentiation (Ying et al. 2011; Garcia et al. 2014). Furthermore, the dietary molecules like curcumin and piperine are obtained from turmeric and black and long peppers, respectively, known for anticancer effects, where they target breast CSCs by inhibiting Wnt/β-catenin and Notch signalling pathways (Kakarala et al. 2010). Akt/mTOR signalling pathway is critical role in the survival and invasion of CSCs. Of note, Akt inhibition is a cause for apoptosis and inhibition of CSC motility (Dandawate et al. 2013). Selenium, an anticancer dietary molecule, acts biologically in the form of selenoproteins, which are known for oxidoreductase scavenging oxidants (Hatfield et al. 2014). It was confirmed that selenium has implications in the regulation of arachidonic acid metabolism, which could induce apoptosis in leukaemia CSCs (Gandhi et al. 2014) by modulated Akt/mTOR signalling pathway (Kim and He 2013). Sulforaphane, a dietary molecule of cruciferous vegetables (broccoli), inhibits the breast and pancreatic CSC viability by suppressing Wnt/β-catenin signalling pathways (Li et al. 2010; Kallifatidis et al. 2009) or by inhibiting hedgehog signalling (Rodova et al. 2012). Series of studies revealed that the sulforaphane inhibits the Akt signalling pathway in variety of solid tumour cancers (Chaudhuri et al. 2007) and also inhibits the CSCs in breast cancer (Korkaya et al. 2009). Owing to ubiquitous occurrences of dietary polyphenols especially flavonoids can be easily made available from a broad range of fruits, grains and vegetables. Plethora of studies has suggested that different extracts containing dietary polyphenols can significantly influence CSCs and cancer cells (Sak and Everaus 2015). Under in vitro experiments, genistein (soya isoflavone) or blueberry polyphenol is reported to reduce the population of breast CSC-like cells (Montales et al. 2012). Interestingly, the cocktail of green tea catechins mixed with sulforaphane and quercetin significantly reduced the viability, migration and induced apoptosis in pancreatic CSCs (Appari et al. 2014). Quercetin a dietary bioactive molecule from apples, teas, onions and berries have considerable anticancer activity against pancreatic and head/neck CSCs (Chang et al. 2013; Zhou et al. 2010). Quercetin in combination with EGCG or sulforaphane has synergistic activity to eliminate the prostate or pancreatic CSCs (Srivastava et al. 2011; Tang et al. 2010). Under clinical trial studies, curcumin demonstrated its therapeutic potential by targeting CSCs (Li and Zhang 2014). The overview of selected signalling pathways implicated in maintenance of cancer stemness, survival, proliferation and differentiation of CSCs, and the targets of dietary substances are summarized in Fig. 9.1.

Conclusion

There is emerging evidence in favour of supporting the beneficial effects of dietary bioactive metabolites against CSCs and cancer cells. The main positive concern of the dietary ingredients is that they are naturally available in edible dietary sources. But there is also need to establish the safety and efficacy of these molecules by performing relevant clinical trials. In addition, the dietary resources can influence CSCs in combination with conventional chemotherapy drugs by inhibiting the key signalling pathways for survival, stemness maintenance of CSCs etc. The synergistic effects may also target the CSCs. However, rigorous clinical trials need to be conducted to address the potential antagonistic effects like counteraction of dietary metabolites against chemotherapy drugs along with the additive toxicity. The circumstantial scientific literature describing the effects of specific diet and dietary metabolites on various targets of cancer progression clearly suggests that the rational selection and daily consumption of foods containing sufficient amounts of the bioactive metabolites described above may be beneficial for the prevention of metastasis and might improve cancer prognosis. Overall, the inclusion of dietary polyphenols (mostly flavonoids) can be considered as promising strategy for targeting CSCs in different cancers; however, further extensive experimentation involving the issues of their bioavailability, molecular mechanisms at the cellular and molecular levels needs to be conducted under preclinical and clinical settings.

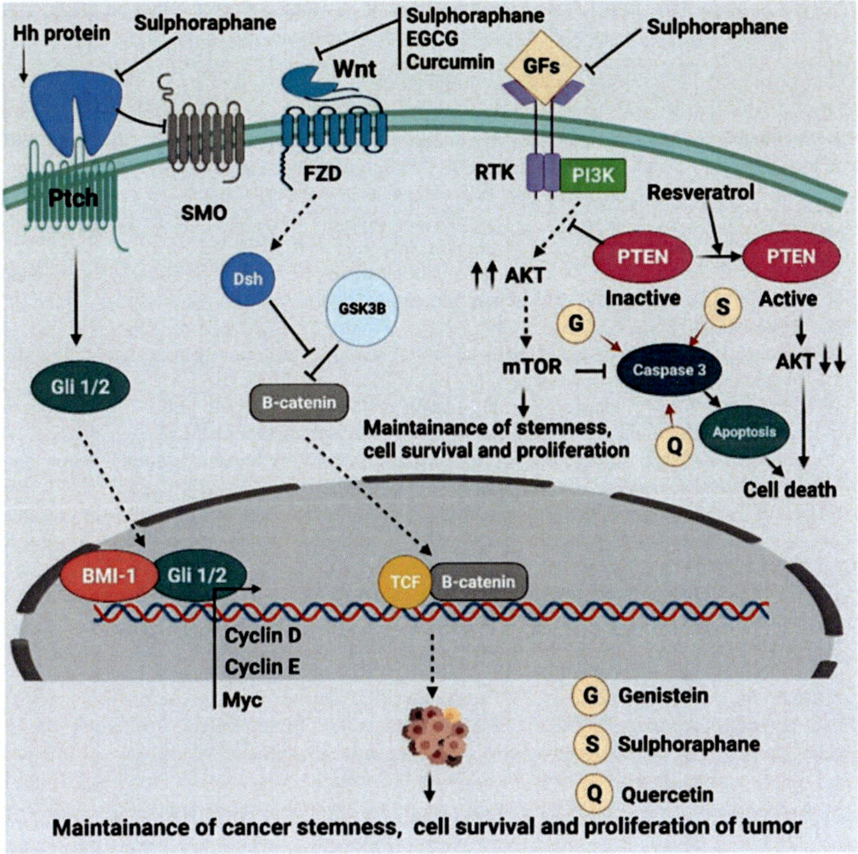

Fig. 9.1 Overview of selected signalling pathways implicated in maintenance of cancer stemness, survival, proliferation and differentiation of CSCs. Most notably, the pathways like Wnt, Notch, PTEN, Hedgehog and PI3K/Akt under normal circumstances are involved in development and maintenance of cancer stemness, survival and thereby proliferation of tumours. These novel regulatory mechanisms and their crosstalk promote self-renewal and differentiation. These pathways have been targeted for tailoring the therapeutic agents against the CSCs. Series of dietary ingredients inhibit the CSC-associated signalling pathways and ultimately adversely affect maintenance of CSC stemness, survival and proliferation. Of note, resveratrol inhibits MTA1/HDAC complex and restores acetylated active PTEN, which downregulates the Akt pathways and leads to cell death. Polyphenols like genistein, sulforaphane and quercetin inhibit the proliferation and/or induce caspase-3-dependent apoptosis of cancer cells

References

Appari M, Babu KR, Kaczorowski A, et al. Sulforaphane, quercetin and catechins complement each other in elimination of advanced pancreatic cancer by miR-let-7 induction and K-ras inhibition. Int J Oncol. 2014;45:1391–400.

Beyaz S, Mana MD, Roper J, Kedrin D, Saadatpour A, Hong S-J, et al. High-fat diet enhances stemness and tumorigenicity of intestinal progenitors. Nature. 2016;531(7592):53–8.
Chan MM, Chen R, Fong D. Targeting cancer stem cells with dietary phytochemical—repositioned drug combinations. Cancer Lett. 2018;433:53–64.
Chang WW, Hu FW, Yu CC, et al. Quercetin in elimination of tumor initiating stem-like and mesenchymal transformation property in head and neck cancer. Head Neck. 2013;35:413–9.
Chaudhuri D, Orsulic S, Ashok BT. Antiproliferative activity of sulforaphane in Akt-overexpressing ovarian cancer cells. Mol Cancer Ther. 2007;6:334–45.
Cojoc M, Mabert K, Muders MH, Dubrovska A. A role for cancer stem cell in therapy resistance: cellular and molecular mechanisms. Semin Cancer Biol. 2015;31:1627.
Creighton CJ, Li X, Landis M, et al. Residual breast cancers after conventional therapy display mesenchymal as well as tumor-initiating features. Proc Natl Acad Sci U S A. 2009;106:13820–5.
Dandawate P, Padhye S, Ahmad A, et al. Novel strategies targeting cancer stem cells through phytochemicals and their analogs. Drug Deliv Transl Res. 2013;3:165–82.
Fornari C, Beccuti M, Lanzardo S, et al. A mathematical-biological joint effort to investigate the tumor-initiating ability of Cancer Stem Cells. PLoS One. 2014;9:e106193.
Gandhi UH, Kaushal N, Hegde S, et al. Selenium suppresses leukemia through the action of endogenous eicosanoids. Cancer Res. 2014;74:3890–901.
Garcia JJ, Lopez-Pingarron L, Almeida-Souza P, et al. Protective effects of melatonin in reducing oxidative stress and in preserving the fluidity of biological membranes: a review. J Pineal Res. 2014;56:225–37.
Garcia-Mayea Y, Mir C, Masson F, Paciucci R, LLeonart ME. Insights into new mechanisms and models of cancer stem cell multidrug resistance. Semin Cancer Biol. 2020;60:166–80.
Gottesman MM, Fojo T, Bates SE. Multidrug resistance in cancer: role of ATP-dependent transporters. Nat Rev Cancer. 2002;2:48–58.
Graham JG, Quinn ML, Fabricant DS, Farnsworth NR. Plants used against cancer—an extension of the work of Jonathan Hartwell. J Ethnopharmacol. 2000;73:347–77.
Hatfield DL, Tsuji PA, Carlson BA, et al. Selenium and selenocysteine: roles in cancer, health, and development. Trends Biochem Sci. 2014;39:112–20.
Hill RP. Identifying cancer stem cells in solid tumors: case not proven. Cancer Res. 2006;66:1891–5.
Jin X, Jin X, Kim H. Cancer stem cells and differentiation therapy. Tumour Biol. 2017;39 (10):1010428317729933.
Jones CL, Stevens BM, D'Alessandro A, Reisz JA, Culp-Hill R, Nemkov T, et al. Inhibition of amino acid metabolism selectively targets human leukemia stem cells. Cancer Cell. 2018;34 (5):724–40.
Kakarala M, Brenner DE, Korkaya H, et al. Targeting breast stem cells with the cancer preventive compounds curcumin and piperine. Breast Cancer Res Treat. 2010;122:777–85.
Kallifatidis G, Rausch V, Baumann B, et al. Sulforaphane targets pancreatic tumour-initiating cells by NF-kappaB-induced antiapoptotic signalling. Gut. 2009;58:949–63.
Kim I, He YY. Targeting the AMP-activated protein kinase for cancer prevention and therapy. Front Oncol. 2013;3:175.
Kim DH, Surh YJ. Chemopreventive and therapeutic potential of phytochemicals targeting cancer stem cells. Curr Pharmacol Rep. 2015;1:302–11.
Kim YS, Farrar W, Colburn NH, Milner JA. Cancer stem cells: potential target for bioactive food components. J Nutr Biochem. 2012;23:691–8.
Korkaya H, Paulson A, Charafe-Jauffret E, et al. Regulation of mammary stem/progenitor cells by PTEN/Akt/beta-catenin signaling. PLoS Biol. 2009;7:e1000121.
Lathia JD, Hitomi M, Gallagher J, et al. Distribution of CD133 reveals glioma stem cells self-renew through symmetric and asymmetric cell divisions. Cell Death Dis. 2011;2:e200.
Le NH, Franken P, Fodde R. Tumour–stroma interactions in colorectal cancer: converging on beta-catenin activation and cancer stemness. Br J Cancer. 2008;98:1886–93.

Lee SH, Nam HJ, Kang HJ, et al. Epigallocatechin-3-gallate attenuates head and neck cancer stem cell traits through suppression of Notch pathway. Eur J Cancer. 2013;49:3210–8.
Li Y, Zhang T. Targeting cancer stem cells by curcumin and clinical applications. Cancer Lett. 2014;346:197–205.
Li Y, Zhang T, Korkaya H, et al. Sulforaphane, a dietary component of broccoli/broccoli sprouts, inhibits breast cancer stem cells. Clin Cancer Res. 2010;16:2580–90.
Li W, Zimmerman SE, Peregrina K, Houston M, Mayoral J, Zhang J, et al. The nutritional environment determines which and how intestinal stem cells contribute to homeostasis and tumorigenesis. Carcinogenesis. 2019;40(8):937–46.
Liao WT, Ye YP, Deng YJ, Bian XW, Ding YQ. Metastatic cancer stem cells: from the concept to therapeutics. Am J Stem Cells. 2014;3(2):46–62.
Liskova A, Kubatka P, Samec M, Zubor P, Mlyncek M, Bielik T, et al. Dietary phytochemicals targeting cancer stem cells. Molecules. 2019;24:899.
Mani SA, et al. The epithelial–mesenchymal transition generates cells with properties of stem cells. Cell. 2008;133:704–15.
Mihaylova MM, Sabatini DM, Yilmaz OH. Dietary and metabolic control of stem cell function in physiology and cancer. Cell Stem Cell. 2014;14:292–305.
Mineva ND, Paulson KE, Naber SP, et al. Epigallocatechin-3-gallate inhibits stem-like inflammatory breast cancer cells. PLoS One. 2013;8:e73464.
Montales MT, Rahal OM, Kang J, et al. Repression of mammosphere formation of human breast cancer cells by soy isoflavone genistein and blueberry polyphenolic acids suggests diet-mediated targeting of cancer stem-like/progenitor cells. Carcinogenesis. 2012;33:652–60.
Nobili S, Lippi D, Witort E, et al. Natural compounds for cancer treatment and prevention. Pharmacol Res. 2009;59:365–78.
Rodova M, Fu J, Watkins DN, et al. Sonic hedgehog signaling inhibition provides opportunities for targeted therapy by sulforaphane in regulating pancreatic cancer stem cell self-renewal. PLoS One. 2012;7:e46083.
Saito Y, Uchida N, Tanaka S, et al. Induction of cell cycle entry eliminates human leukemia stem cells in a mouse model of AML. Nat Biotechnol. 2010;28:275–80.
Sak K, Everaus H. Role of flavonoids in future anticancer therapy by eliminating the cancer stem cells. Curr Stem Cell Res Ther. 2015;10:271–82.
Srivastava RK, Tang SN, Zhu W, et al. Sulforaphane synergizes with quercetin to inhibit self-renewal capacity of pancreatic cancer stem cells. Front Biosci. 2011;3:515–28.
Tang DG. Understanding cancer stem cell heterogeneity and plasticity. Cell Res. 2012;22:457–72.
Tang SN, Singh C, Nall D, et al. The dietary bioflavonoid quercetin synergizes with epigallocathechin gallate (EGCG) to inhibit prostate cancer stem cell characteristics, invasion, migration and epithelial-mesenchymal transition. J Mol Signal. 2010;5:14.
Visvader JE, Lindeman GJ. Cancer stem cells in solid tumours: accumulating evidence and unresolved questions. Nat Rev Cancer. 2008;8:755–68.
Yang L, Shi P, Zhao G, et al. Targeting cancer stem cell pathways for cancer therapy. Signal Transduct Target Ther. 2020;5(1):8.
Ying M, Wang S, Sang Y, et al. Regulation of glioblastoma stem cells by retinoic acid: role for Notch pathway inhibition. Oncogene. 2011;30:3454–67.
Zhou W, Kallifatidis G, Baumann B, et al. Dietary polyphenol quercetin targets pancreatic cancer stem cells. Int J Oncol. 2010;37:551–61.

Chapter 10
Interaction of Dietary Metabolites with Anticancer Drug Targets

Abstract A vast body of literature has accumulated in the past two decades, which is consistently reporting the significance of dietary bioactive metabolites as possible target-oriented anticancer agents. Diet origin substances like curcumin, sulforaphane, resveratrol, quercetin, epigallocatechin gallate and hundreds of dietary polyphenols have been extensively studied as anticancer agents in preclinical and clinical settings. Until now, over hundreds of different dietary molecules have been identified, which need to be translated into clinical practice so as to expand their clinical utility in medicine as an adjuvant therapy. Employing approaches like proteomics, transcriptomics, metabolomics and microbiomics might pave the way for understanding of how dietary factors interact with hallmarks of cancer drug targets and chemotherapy drugs and contribute in cancer prevention. Nevertheless, there is also an equal research opportunity to conduct research for improving the bioavailability of the effective anticancer dietary agents.

Keywords Dietary bioactive molecules · Anticancer drug targets · Growth · Inhibition · Signalling pathways · Growth factors

Abbreviations

AKT	Protein kinase B
AMPK	AMP-activated protein kinase
AP-1	Activator protein 1
Bax	Bcl-2-associated X protein
Bcl-2	B-cell lymphoma 2
Bcl-xL	B-cell lymphoma extra large,
BMP2	Bone morphogenetic protein 2
Caspase	Cysteine proteases with aspartate specificity
CDKs	Cyclin-dependent kinases
COX-2	Cyclooxygenase-2
CXCR-4	C-X-C chemokine receptor type 4
NOX	NADPH oxidase

R. N. Gacche, *Dietary Research and Cancer*,
https://doi.org/10.1007/978-981-16-6050-4_10

E2F1	E2F Transcription factor 1
E-cadherin	Epithelial cadherin
EGCG	Epigallocatechin-3-gallate
EGFR	Epidermal growth factor receptor
EMT	Epithelial-to-mesenchymal transition
Erk	Extracellular signal-regulated kinase
ERα	Oestrogen receptor alpha
FAK	Focal adhesion kinase
FGF	Fibroblast growth factor
FOS	Fos proto-oncogene
FOXO	Forkhead/winged-helix box gene, group O
GM-CSF	Granulocyte–macrophage colony-stimulating factor
GST	Glutathione S-transferase
HER2	Human epidermal growth factor receptor 2
HGF	Hepatocyte growth factor
HMC	Hallmark of cancer
ICAM-1	Intercellular adhesion molecule 1
IGF	Insulin-like growth factor
IGF1-R	Insulin-like growth factor 1 receptor
IGFR	Insulin-like growth factor receptor
IKK	IκB kinase
Il-6	Interleukin-6
IL-8	Interleukin-8
IRE-1	Inositol-requiring enzyme 1
Jak	Janus family tyrosine kinase
JNK	c-Jun N-terminal kinase
KDR	Kinase insert domain receptor
LOX	Lysyl oxidase
MAPK	Mitogen-activated protein kinase
MDM-2	Murine double minute 2, a p53-associated oncogene
MEK	MAPK/Erk kinase
miRNA	microRNA
MMP	Matrix metalloproteinase
mTOR	Mammalian target of rapamycin
NDPK	Nucleoside diphosphate kinase
NF-kB	Nuclear factor κ B
Nrf2	Nuclear factor erythroid 2
ODC	Ornithine decarboxylase
p53	Tumour suppressor protein that protects from DNA damage
P70S6K	Ribosomal protein S6 kinase
PAK	p21-activating kinases
PARP	Poly (ADP-ribose) polymerase
PDGFR	Platelet-derived growth factor receptor
PI3K	Phosphoinositide 3 kinase

PIP3	Prolactin-induced protein-3
PKC	Protein kinase C
PPAR-γ	Peroxisome proliferator activator receptor-gamma
QR	Quinone oxidoreductase
Raf	Rapidly accelerated fibrosarcoma (RAF kinases)
RAS	Rat sarcoma
RASD1	Dexamethasone-induced Ras-related protein 1
ROS	Reactive oxygen species
SDF-1	Stromal cell-derived factor-1
SKP2	S-phase kinase-associated protein 2
SNAI1	Snail family zinc finger 1,
Src	Rous sarcoma oncogene cellular homolog
STAT3	Signal transducer and activator of transcription 3
TBK1	TANK-binding kinase 1
TGF-β	Transforming growth factor β
TNF-α	Tumour necrosis factor-α
TP53	Tumour protein p53
VEGF	Vascular endothelial factor
VEGFR	VEGF receptor
ZEB1	Zinc finger E-box-binding homeobox 1

Introduction

While describing the molecular mechanisms of dietary interactions with evolving anticancer drug targets, it is imperative to consider the framework of molecular interactions that help to protect from cancer. The rational framework is to investigate the dietary impact and interactions with the eight HMC (Hanahan and Weinberg 2011). One more aspect of addressing the dietary interactions is the impact of diet on chemotherapy drugs, which influence their pharmacokinetic or pharmacodynamic properties and thereby affect their therapeutic index (Segal et al. 2014). In general, variety of dietary agents have been described to regulate variety of cancer-associated cell signalling pathways, for example, series of growth factors receptors such as EGFR, VEGFR and IGF1-R; several protein kinases like Ras/Raf, mTOR, PI3K, Bcr-abl and AMPK; series of transcription factors such as nuclear factor κB (NF-κB), activator protein-1 (AP-1), signal transducer and activator of transcription 3 (STAT3), Sonic Hedgehog, β-catenin/Wnt, nuclear factor erythroid 2 (Nrf2), peroxisome proliferator activator receptor-gamma (PPAR-γ) and few pro-inflammatory molecules like TNF-α, interleukins, 5-LOX and COX-2. Dietary bioactive substances are also known to induce apoptosis in variety of cancers and interact with apoptosis-related proteins such as caspases, Bcl-2, Bax, Bad and PARP (Pandey et al. 2017). Plethora of literature accumulated in the recent past describing the target-oriented anticancer properties of dietary bioactive molecules has been extensively reviewed, and the comprehensive profile of dietary anticancer agents, their dietary sources, chemical structures and anticancer drug targets are summarized in Table 10.1 (Aggarwal and Shishodia 2006; Pandey et al. 2017; Sanchez-Vega et al. 2018; Nosrati et al. 2017; Logan and Bourassa 2018; Chae et al. 2019; Ahmad et al. 2020).

Of the eight HMCs, the first is related to sustained proliferative cell signalling pathways supporting the cancer growth and proliferation. The cancer genome atlas describes that series of transcription factors like nuclear factor-kappa B (NF-κB), signal transducer and activator of transcription family, activator protein 1 (AP-1), β-catenin/Wnt, PPAR-γ, etc., growth factors and receptors such as epidermal growth factor receptor (EGFR), vascular endothelial growth factor receptor (VEGFR), insulin-like growth factor (IGF) 1-receptor; several protein kinases like Ras/Raf mTOR, PI3K/AKT, Bcr-abl, AMPK, variety of pro-inflammatory cytokines like tumour necrosis factor-α (TNF-α), interleukins, pro-inflammatory enzymes like cyclooxygenase (COX)-2, 5-LOX, proteasome and other epigenetic changes collectively drive the cancer progression by coordination and crosstalk with variety of factors and signalling pathways (Sanchez-Vega et al. 2018).

Dietary Metabolites as Inhibitors of Variety of Anticancer Targets

The research findings are consistently evolving and linking the role of dietary metabolites as inhibitors of variety of anticancer targets. For example, series of dietary bioactive molecules like berberine (oregon grape, barberry and goldenseal), cardamonin (black cardamom), capsaicin (peppers), ursolic acid (apples, cranberries, peppermint, rosemary), 6-gingerol (ginger), butein (beetroots), sulforaphane and indole-3-carbinol (cruciferous vegetables), dibenzoylmethane (licorice), diosgenin (fenugreek), gambogic acid (kokum), thymoquinone (black cumin), xanthohumol (hop), zerumbone (pinecone ginger), natural antioxidant mixture (spinach), beta-carotenoids (carrots), ellagic acid (pomegranate), myricetin (cranberry), carnosol (rosemary), vanillin (vanilla), eugenol (cloves), vitamin C (citrus fruits), resveratrol (red grapes and peanuts), allicin (garlic) and lycopene (tomato) have potentials to inhibit the cancer cell growth by targeting NF-κB, PPAR-γ, β-catenin/Wnt signalling and majority of them activate caspase-mediated apoptosis pathway in different cancer cells in several preclinical model studies. The cancer cell signalling pathways like mTOR is inhibited by curcumin, diosgenin and ursolic acid, while eugenol, cardamonin, zerumbone, thymoquinone and xanthohumol demonstrated inhibitory activity against TNF-α expression. Dietary molecules are also reported to inhibit several anticancer drug targets like NRF2, hedgehog signalling pathways, EGFR, VEGFR, IGFR, PDGFR, Ras/Raf, PI3K, AMPK, BCR-ABL, TNF, interleukins, COX-2, LOX and MMP-9 in preclinical model studies, and several food ingredients like capsaicin, curcumin, ginger and anethole have also been tested under clinical

Table 10.1 Profile of dietary anticancer agents, their dietary sources, chemical structures and anticancer drug targets

Name of the dietary anticancer agent	Dietary sources	Chemical structure	Anticancer targets
Acetoxychavicol acetate	*Alpinia galanga* (blue ginger)	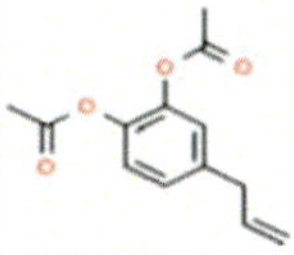	c-Myc, survivin, XIAP Bcl-2, Bcl-xL, Bfl-1/A1, cyclin-D1, FLIP, IAP-1/-2
Allicin	*Allium sativum* (garlic)	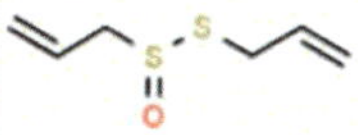	FGF2, ICAM-1, VEGF
Anacardic acid	*Semecarpus anacardium* (cashew)	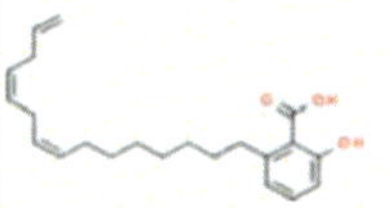	Bcl-2, Bcl-xL, c-FLIP, cIAP-1, COX-2, cyclin-D1, survivin
Anethole	Pimpinella anisum (anise)	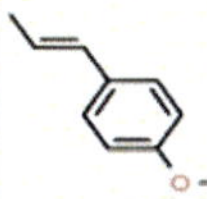	AKT, AP-1, AR, ER, JAK, MAPK, Keap1, NF-κB, Nrf2, Notch, p53, STAT, Wnt
Apigenin	Fresh parsley, vine spinach, celery seed, green celery heart, Chinese celery, red and white sorghum, oranges, kumquats, onions, wheat sprouts, tea and cilantro	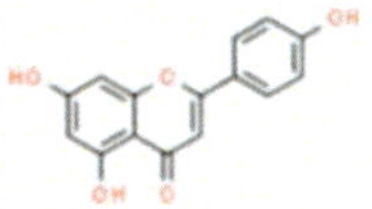	HIF-1, ICAM-1, VEGF, JAK2/STAT3, p53/p21Cip1, Nrf2
Baicalein	Olive leaves (*Scutellaria galericulata*)	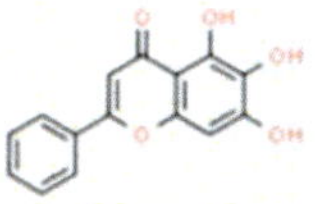	Rb/E2F/cyclin-CDK4/p53, PI3K/AKT, MAPKs
Berberine	*Berberis vulgaris* (barberry), Oregon grape and tree turmeric	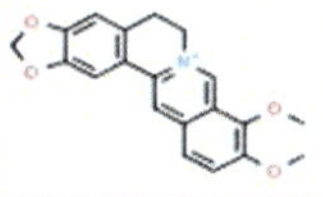	Bcl-2, Bcl-xL, cyclin-D1, c-Myc, FAK, IKK, MMP-2/-9, NF-κB, u-PA
Butein	Cashew, pod vegetables, broad bean pod	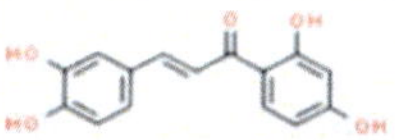	Bcl-2, Bcl-xL, c-Myc, cyclin-D1, ERK-1/-2, IAP-2, NF-κB
Caffeic acid	Coffee, apples, artichoke, berries, olives, pears and wine	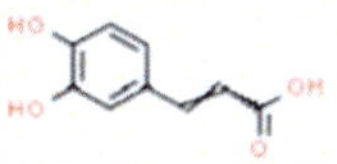	MMP-9, AKT, mTOR, ERK1/2, CDK2
Capsaicin	*Capsicum annuum* (chilli pepper)	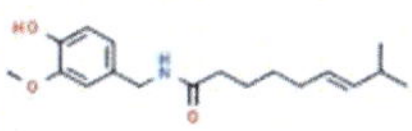	AKT, Bcl-2, Bcl-xL, E2F, p38MAPK, p125(FAK), Rac1, survivin, VEGF

(continued)

Table 10.1 (continued)

Name of the dietary anticancer agent	Dietary sources	Chemical structure	Anticancer targets
Chrysin	Passion fruit, honey and propolis	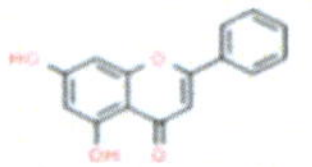	NF-kB, PXR, TBK1, AMPK, AKT, ERK, PPAR
Curcumin	*Curcuma longa* (turmeric)	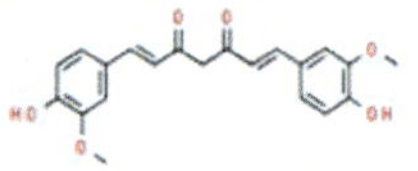	AP-1, Bcl-2, cyclin-D1, IAP-1, NF-κB, PAK1, survivin, VEGF
Diallyl sulphide	*Allium schoenoprasum* (chives)		p53, Bax, Bcl-2, IL-6, IL-1β, MMP-2, MMP-9, TNF-α, VEGF survivin
Daidzein	Soya beans, tofu, soya milk, soya flour, textured soya protein, soya protein isolates, tempeh	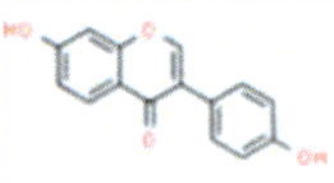	Akt/FOXO3a, NF-κB, Raf/MEK/ERK
Diosgenin	Fenugreek, yam tubers and legumes	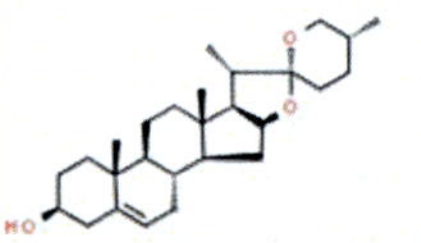	TNF-α, NF-κB, AKT, mTOR, JNK, STAT3, p38/MAPK, c-SRC, JAK1
Diosmin	Citrus fruits	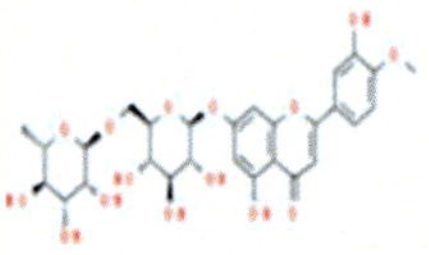	PI3K/AKT/mTOR/NF-κB, TNF-α, COX-2, Bcl-2, Bcl-xL, Mcl-1, Bax, Bad, caspase-3, and caspase-9, MDM2
Epigallocatechin gallate or EGCG	Green tea, white, oolong and black teas, cranberries, strawberries, blackberries, kiwis, cherries, pears, peaches, apples and avocados, pecans, pistachios and hazelnuts	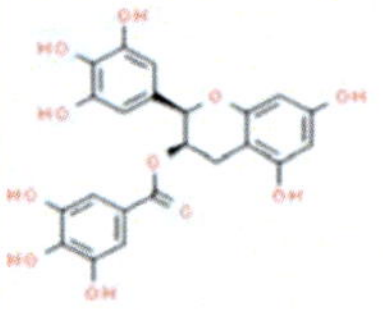	AKT, AP-1, Bcl-2, Bcl-xL, ERK-1/-2, JAK, Mcl-1, NF-κB, PI3K, Ras, STAT, u-PA, VEGF
Ellagic acid	White oak, strawberries, raspberries, blackberries, cherries and walnuts	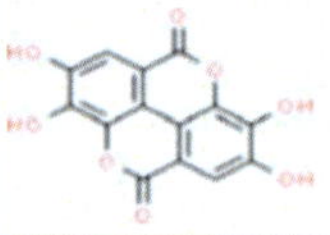	Bcl-2, cyclin D1, cyclin G1, FoxO1, FoxO3a, ERα, NDPK, RASD1
Eugenol	Clove, cinnamon bark and leaves, tulsi turmeric, pepper, ginger, oregano and thyme	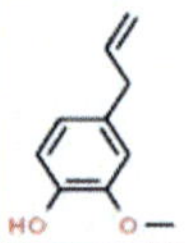	Cyclin D1, E2F1, NF-κB, p21, surviving

(continued)

Table 10.1 (continued)

Name of the dietary anticancer agent	Dietary sources	Chemical structure	Anticancer targets
Fisetin	Strawberries, apples, mangoes, persimmons, kiwis, grapes, tomatoes, onions and cucumbers, nuts and wine	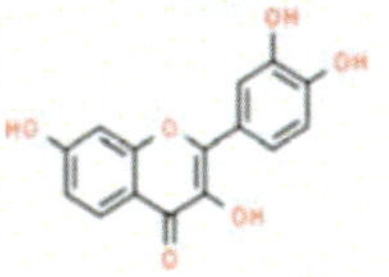	COX-2, EGFR, ERK-1/-2, MMP-2, NF-κB, TAK-1, u-PA, Wnt
Galangin	Medicinal herbs, Hance, Alnus pendula, Matsum, Plantago major, Scutellaria galericulata and propolis	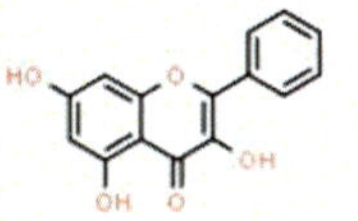	Caspase-3 and 7, p53, Bax, DR5, PFT-α), p21, Akt, p70S6K, PKCα, PKCδ, ERK1/2, phospho-IκBα, c-Fos, c-Jun, NF-κB, AP-1, MMP-2, MMP-9
Garcinol	*Garcinia indica* (kokum)	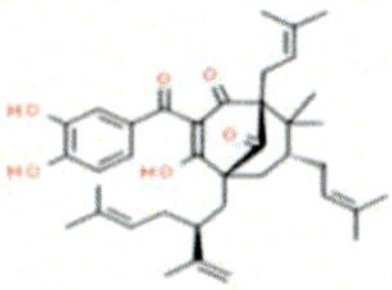	AKT, cyclin-D3, ERK, nicotinic receptor, Src
Genistein	*Glycine max* (soya beans)	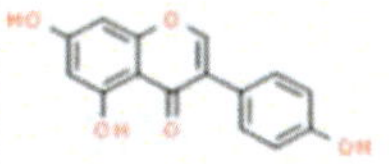	AKT, AP-1, ATM, Bcl-2, Bcl-xL, Chk-1/-2, cdc25, FGF-2, NF-κB, u-PA, VEGF
Gingerol	*Zingiber officinale* (ginger)	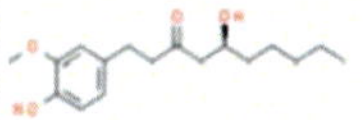	Bcl-2, IL-1β, IL-12, MMP-2, MMP-9, TNF-α, VEGF
Hesperidin	Sweet orange, tangerine, sour orange, lime, lemon, grapefruit	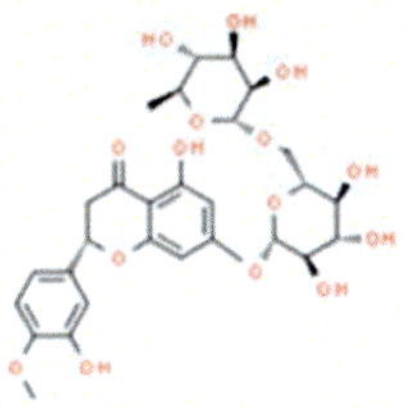	p53, Bax, caspases-3, SDF-1/CXCR-4, PI3K/AK, Nrf2/ARE/HO-1
Indole-3-carbinol	Broccoli, Brussels sprouts, cabbage, collards, cauliflower, kale, mustard greens, turnips and rutabagas	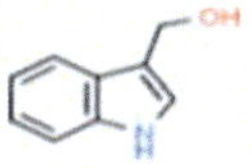	α-, β-, and γ-catenin, caspase-8, E-cadherin, p53
Kaempferol	Apples, broccoli, Chinese cabbage, kale, leeks, lettuce, onions, spinach, chives, dill, fennel leaves, blueberry, cherry, cranberry	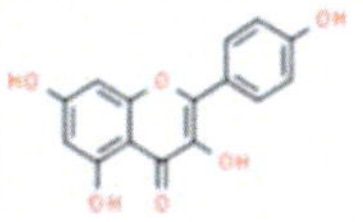	PI3K/AKT, EGFR/ERK, ERRα/γ, IRE1-JNK-CHOP

(continued)

Table 10.1 (continued)

Name of the dietary anticancer agent	Dietary sources	Chemical structure	Anticancer targets
Lupeol	Mango, white cabbage, pepper, cucumber, tomato, in fruits such as olive, fig, strawberry, red grapes	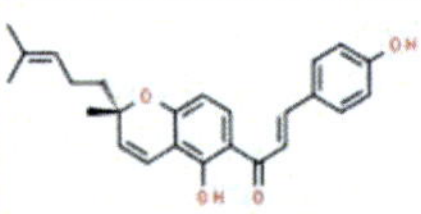	Bax, caspase-3, caspase-9, c-FLIP, NF-κB, survivin
Luteolin	Clover, celery, parsley, broccoli, onion leaves, carrots, peppers, cabbages, apple skins, spinach, beet, cauliflower, artichoke, olive oil, tea, perilla leaf, peppermint, thyme, sage, rosemary and oregano	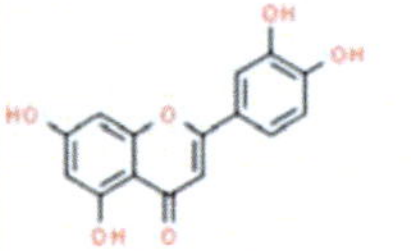	AKT, HDAC8, miR-301, Nrf2, ROS/ER stress
Lycopene	Tomato, watermelon, pink grapefruit, pink guava and papaya	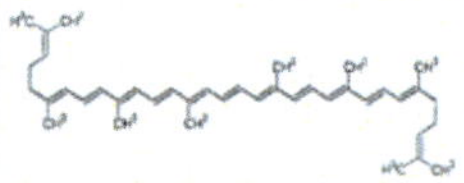	FoxO1, EGFR, p-S6, Skp2
Morin	Tea, coffee, cereals, fruits and red wine	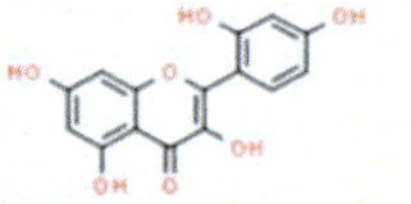	AKT, caspases, p21, STAT3
Myricetin	Cranberry, dock, sweet potato leaves, chard, broad beans, rutabagas, garlic, blueberry, peppers, hot chilli, blackberry, lemon	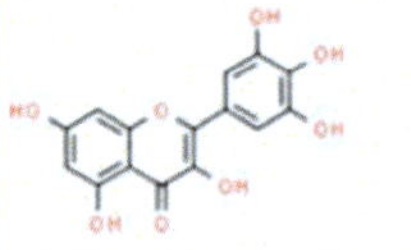	NDPK, mTOR/AKT, BMP-2/Smad, MAPKs, PIM1/CXCR4, Wnt/β-catenin
Naringin	Grapefruits, the peels of orange, grapefruits, including albedo and flavedo	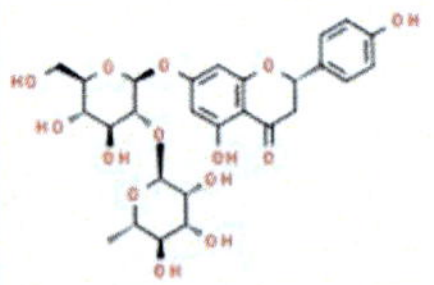	AKT/mTOR, EGFR, NEU3, HER2, Smad3/Smad7, MAPKs, Zeb1, ERRα/VEGF/KDR, ATF3
Naringenin	Citrus fruits, grapefruits, orange fruits, tomato	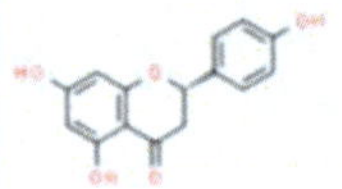	UDP-UGTs, QR, GST, CYP19, ROS, ODC, PTK, PKC, PIP3, G and S phases, CDK, BAX, BCL2, ERα, phosphoinositide, 3-kinase, AKT, ER-α-dependent p38 kinase

(continued)

Table 10.1 (continued)

Name of the dietary anticancer agent	Dietary sources	Chemical structure	Anticancer targets
Nobiletin	Citrus flavonoid extracted from citrus fruits and peels	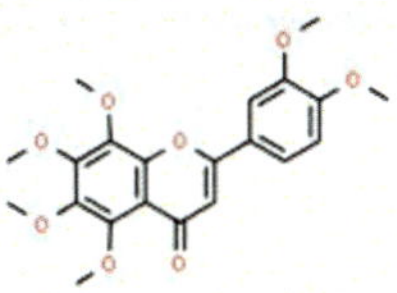	EMT, TGF-β, ZEB, Slug, Snail, STAT3, NF-κB, Akt, PI3K, Wnt, microRNA-7 and microRNA-200b
Phytic acid	Edible seeds, grains, legumes and nuts	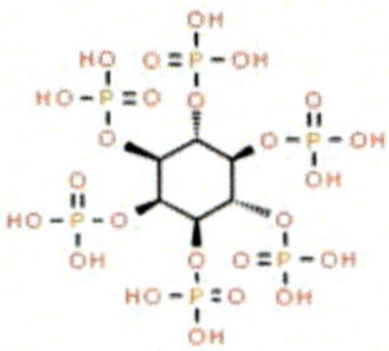	IL-6, IL-8, IL-1β, MMP-9, p21, p53, VEGF
Piperine	*Piper nigrum* (black pepper)	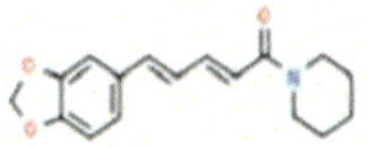	GM-CSF, IL-6, IL-1β, MMP, TNF-α, IL-12p40
Piperlongumine	*Piper longum* (long pepper)	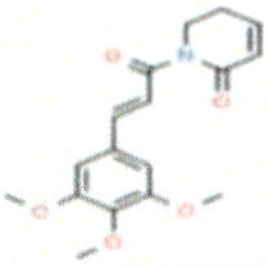	AKT, ERK, MAPK, mTOR, NF-κB, STAT3
Quercetin	Apples, berries, Brassica vegetables, capers, grapes, onions, shallots, tea, tomatoes, parsley	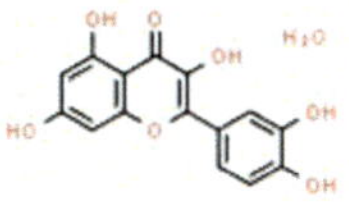	MMP-2/-9, STAT-3, p53, VEGF, BCL2/BAX, AMPK/COX-2, EGFR/VEGFR2, β-catenin, Bcl-xL, cyclin-D1
Resveratrol	Red grapes, peanuts, pistachios, grapes, red and white wine, blueberries, cranberries, and even cocoa and dark chocolate	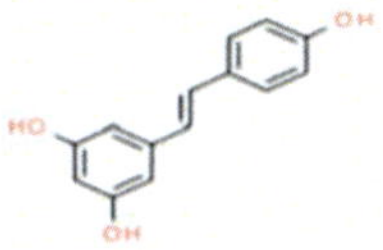	FGF, MAPK, MMP-2/-9, NF-κB, survivin, VEGF
Rosmarinic acid	Rosemary, sage, Spanish sage, basil, oregano	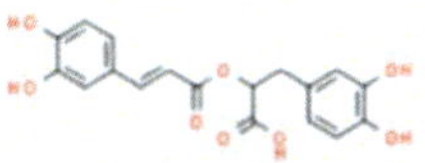	Bcl-2, cyclin D3, IL-8, p21, p27, VEGF
Rutin	Buckwheat, asparagus, unpeeled apples, figs, black tea, green tea and elderflower tea	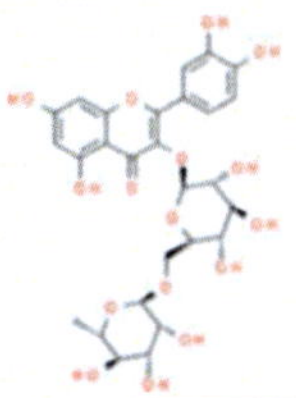	NF-κB, TNF-α, ILs, Beclin, Bcl-2), BAX, caspases, VEGF

(continued)

Table 10.1 (continued)

Name of the dietary anticancer agent	Dietary sources	Chemical structure	Anticancer targets
Sulforaphane	*Brassica oleracea* (broccoli)		Rb-E2F-1, MMPs
Tangeretin	Abundant in the peel citrus species, sweet orange, mandarin	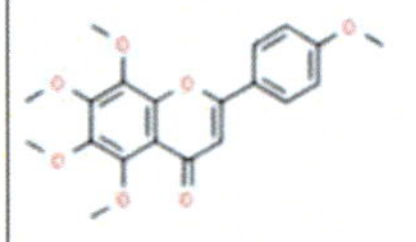	Cell cycle, PI3K/AKT/mTOR, Notch, MAPK, induces autophagic cell death, suppresses migration, invasion, and angiogenesis
Ursolic acid	Rosemary, fruits (apple fruit peel), flowers, berries	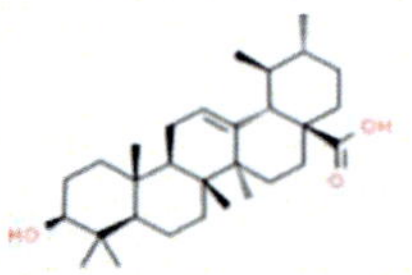	AKT, COX-2, cyclin-D1, JNK, MMP-9, NF-κB, VEGF
Vanillin	*Vanilla planifolia* (vanilla), Vanilla ice cream	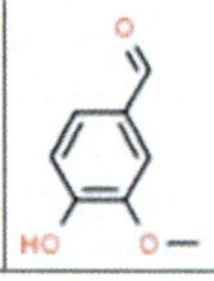	AKT, HGF, MMP-9, PI3K, VEGF

trials (Tan et al. 2011; Sung et al. 2012; Khuda-Bukhsh et al. 2014; Pandey et al. 2017; Nosrati et al. 2017; Logan and Bourassa 2018).

Enabling replicative immortality is one of the HMCs, and plethora of research findings has investigated the role of dietary bioactive metabolites as effective agents in arresting the cell cycle of variety of cancer cells at various phases of cell division. For example, variety of dark coloured berries induce apoptosis by arresting cell cycle arrest at G1 phase via induction of WAF1/p21 and inhibition of cyclins and cyclin-dependent kinases (CDKs) like cdk4, cdk6, cyclin D1 and cyclin D3. Citrus flavonoids and limonoids are known to arrest the cell cycle at S and G2/M phases and induce apoptosis. Series of dietary polyphenols like quercetin, genistein, kaempferol, luteolin, daidzein, apigenin and epigallocatechin have been shown to arrest cell cycle by regulating the activity of the CDKs. Cruciferous vegetables like broccoli and cabbage are rich in phenethyl isothiocyanate and arrest cell cycle at G2/M phase and induce apoptosis in variety of cancer cells. In brief, cell cycle control by dietary agents is one of the bases for cancer chemoprevention (Meeran and Katiyar 2008; Nosrati et al. 2017). The limitations of aforementioned studies lie in their preclinical nature; perhaps clinical settings with impact of dietary agents on cell cycle regulation might explore the medicinal potential of these dietary metabolites for cancer prevention.

Being an HMC, tumour angiogenesis has remained an active research window in the mainstream of anticancer drug development. Angiogenesis is the formation of new blood vessel from existing ones. The pioneering work of Judah Folkman demonstrated that tumour growth requires sustained angiogenesis to retrieve nutrients and oxygen, arresting the tumour angiogenesis may starve the tumour cells and may act as a checkpoint on cancer progression. Since then, the quest for designing anti-angiogenic agents accelerated in logarithm fashion and culminated in the development of 11 FDA anti-angiogenic drugs (Gacche 2015; Gacche and Assaraf 2018). Of note, until the end of November 2019, there were over 3739 registered clinical trials at Clinical trials.gov when searched with keyword "Angiogenesis inhibitors". Majority of the study designs are related to investigating the single or combinatorial effect of FDA-approved drugs on angiogenesis in different types of cancer patients. Specifically, there are five clinical trials (NCT02969070; NCT00296595; NCT00654459; NCT01085019; and NCT00166088) that have been conducted to assess the impact of dietary nutraceuticals on endothelial function and dysfunction. The outcome of such trials can be utilized as a resource for designing RCTs as endothelial cells happen to be a building block of vascular network. Nevertheless, the majority of the FDA-approved anticancer drugs are targeted towards endothelial cells (Gacche and Meshram 2014).

A vast body of preclinical and clinical data has accumulated in the recent years addressing the anti-angiogenic effects of dietary sources and provides inspiration for formulating "anti-angiogenic foods", which perhaps can be an alternative therapeutic approach for targeting tumour angiogenesis and thereby cancer prevention. Variety of fruits like strawberries, raspberries, blackberries, apple, blueberries, cranberries, pineapple, cherries, grapefruit, oranges, lemons, red grapes, green tea, pomegranate and series of bioactive molecules such as carotenoids (lycopene), flavonoids (in particular glycosides), anthocyanidins (anthocyanins), flavanones (hesperetin) and stilbenoids (resveratrol) have demonstrated anti-angiogenic effects in variety of cancer cells. Moreover, different vegetables and mushrooms, soya beans, tomatoes, turmeric, ginger, nutmeg, cinnamon, red propolis, artichokes, garlic, kale, Brussels sprouts, extra-virgin olive oil, grapeseed oil, broccoli, cauliflower, bok choy, lavender, parsley, maitake mushrooms and pumpkins and series of bioactive metabolites like curcumin, flavones (apigenin), isoflavones (genistein), flavonols (quercetin), isothiocyanate (sulforaphane), oleic acid, phenylethanoids and variety of dietary polyphenols present therein have impressive effects on perturbing the vascular organization (Gacche et al. 2011, 2015; Cerezo et al. 2015; Morbidelli et al. 2018; Diniz et al. 2017). Over 17 clinical trials have been conducted involving the implications of dietary substances on diseases characterized by aberrant angiogenesis (Morbidelli et al. 2018). Mediterranean dietary pattern has been recommended as an effective anti-angiogenic food owing to the presence of several bioactive constituents, which have impressive anti-angiogenic potential (Martínez-Poveda et al. 2019). An interesting preclinical animal model study has demonstrated that a moderate dietary restriction of 30–40% has significant impact on disruption of angiogenesis in mouse brain tumour (Mukherjee et al. 2002).

Conclusion

Until now, hundreds of dietary bioactive molecules have been identified as potent anticancer agents, which need to translate into clinical practice so as to expand their utility in medicine as an adjuvant therapy (Pandey et al. 2017). In fact, the endeavours of understanding the mechanisms of action of dietary constituents in cancer epidemiology are associated with complexity as the impact of dietary intervention is equally interlinked with range of food exposures, diversity of the biological targets, interacting metabolic pathways, exposures encountered by the individuals over his or her life period, for example, the state of metabolism, hormones, immune surveillance, physical activity, body composition, inflammation, the dynamics of gut microbiome, oxidative stress, diet and alcohol, ageing and the state of psychosocial and environmental exposures. The entire imbroglio is interlinked and dynamic, making the measurement and analysis of dietary impact highly complex. To circumvent these complexities, the data-driven "omics" approaches like metabolomics, proteomics, transcriptomics and microbiomics might pave the way for understanding of how dietary factors interact with drug targets and chemotherapy drugs and contribute in cancer prevention (Wild 2012).

References

Aggarwal BB, Shishodia S. Molecular targets of dietary agents for prevention and therapy of cancer. Biochem Pharmacol. 2006;71(10):1397–421.

Ahmad R, Khan MA, Srivastava AN, Gupta A, Srivastava A, Jafri TR, Siddiqui Z, Chaubey S, Khan T, Srivastava AK. Anticancer potential of dietary natural products: a comprehensive review. Anti Cancer Agents Med Chem. 2020;20(2):122–236.

Cerezo AB, Winterbone MS, Moyle CW, Needs PW, Kroon PA. Molecular structure-function relationship of dietary polyphenols for inhibiting VEGF-induced VEGFR-2 activity. Mol Nutr Food Res. 2015;59:2119–31.

Chae HS, Xu R, Won JY, Chin YW, Yim H. Molecular targets of genistein and its related flavonoids to exert anticancer effects. Int J Mol Sci. 2019;20(10):2420.

Diniz C, Suliburska J, Ferreira IMPLVO. New insights into the antiangiogenic and proangiogenic properties of dietary polyphenols. Mol Nutr Food Res. 2017;61(6).

Gacche RN. Compensatory angiogenesis and tumor refractoriness. Oncogenesis. 2015;4:e153.

Gacche RN, Assaraf YG. Redundant angiogenic signaling and tumor drug resistance. Drug Resist Updat. 2018;36:47–76.

Gacche RN, Meshram RJ. Angiogenic factors as potential drug target: efficacy and limitations of anti-angiogenic therapy. Biochim Biophys Acta. 2014;1846(1):161–79.

Gacche RN, Shegokar HD, Gond DS, Yang Z, Jadhav AD. Evaluation of selected flavonoids as antiangiogenic, anticancer, and radical scavenging agents: an experimental and in silico analysis. Cell Biochem Biophys. 2011;61(3):651–63.

Gacche RN, Meshram RJ, Shegokar HD, More R, Gond DS, Patil K. Flavanoids as a scaffold for development of novel anti-angiogenic agent: an experimental and computational enquiry. Arch Biochem Biophys. 2015;577–578:35–48.

Hanahan D, Weinberg RA. Hallmarks of cancer: the next generation. Cell. 2011;144:646–74.

Khuda-Bukhsh AR, Das S, Saha SK. Molecular approaches toward targeted cancer prevention with some food plants and their products: inflammatory and other signal pathways. Nutr Cancer. 2014;66(2):194–205.

Logan J, Bourassa MW. The rationale for a role for diet and nutrition in the prevention and treatment of cancer. Eur J Cancer Prev. 2018;27(4):406–10.

Martínez-Poveda B, Torres-Vargas JA, Ocaña MDC, García-Caballero M, Medina MÁ, Quesada AR. The Mediterranean diet, a rich source of angiopreventive compounds in cancer. Nutrients. 2019;11(9). pii: E2036.

Meeran SM, Katiyar SK. Cell cycle control as a basis for cancer chemoprevention through dietary agents. Front Biosci. 2008;13:2191–202.

Morbidelli L, Terzuoli E, Donnini S. Use of nutraceuticals in angiogenesis-dependent disorders. Molecules. 2018;23(10). pii: E2676.

Mukherjee P, El-Abbadi MM, Kasperzyk JL, Ranes MK, Seyfried TN. Dietary restriction reduces angiogenesis and growth in an orthotopic mouse brain tumour model. Br J Cancer. 2002;86 (10):1615–21.

Nosrati N, Bakovic M, Paliyath G. Molecular mechanisms and pathways as targets for cancer prevention and progression with dietary compounds. Int J Mol Sci. 2017;18:2050.

Pandey MK, Gupta SC, Nabavizadeh A, Aggarwal BB. Regulation of cell signaling pathways by dietary agents for cancer prevention and treatment. Semin Cancer Biol. 2017;46:158–81.

Sanchez-Vega F, Mina M, Armenia J, Chatila WK, Luna A, La KC. Oncogenic signaling pathways in the cancer genome atlas. Cell. 2018;173:321–37.

Segal EM, Flood MR, Mancini RS, Whiteman RT, Friedt GA, Kramer AR, Hofstetter MA. Oral chemotherapy food and drug interactions: a comprehensive review of the literature. J Oncol Pract. 2014;10(4):e255–68.

Sung B, Prasad S, Yadav VR, Aggarwal BB. Cancer cell signalling pathways targeted by spice-derived nutraceuticals. Nutr Cancer. 2012;64(2):173–97.

Tan AC, Konczak I, Sze DMY, Ramzan I. Molecular pathways for cancer chemoprevention by dietary phytochemicals. Nutr Cancer. 2011;63(4):495–505.

Wild CP. The exposome: from concept to utility. Int J Epidemiol. 2012;41:24–32.

Chapter 11
Revisiting the Anticancer Drug–Food Interactions

Abstract Chemotherapy anticancer drugs and plant-based foods (mostly fruits, vegetables, whole grains, fruit juices and also derived beverages) have an intricate association with human health and cancer progression. The negative part of food–drug interactions is associated with alterations in the therapeutic effects of drug molecules by affecting their pharmacokinetic or pharmacodynamic properties leading to adverse clinical outcomes. This aspect is an important framework that needs to be considered because diet–drug interactions are likely to happen during the episodes of chemotherapy treatment. Taking into consideration the public health significance of the issue, the US Food and Drug Administration (fda.gov) has provided "Clinical Drug Interaction Studies-Cytochrome P450 Enzyme- and Transporter-Mediated Drug Interactions Guidance for Industry". The circumstantial literature describes that oral chemotherapy drugs and food interactions are associated with significant impact on the pharmacokinetics of orally administered anticancer agents, which adversely affect the bioavailability, absorption and other pharmacokinetic parameters and possibly limit the benefits of chemotherapy drugs. Several herbal products such as garlic, echinacea, ginseng, milk thistle, grapefruit juice and St John's work have demonstrated interactions with variety of cancer chemotherapeutic drugs in humans. Therefore, in the midst of evolving targeted anticancer drug developments it is rational to revisit the complexities of food–drug interaction so as to aware the cancer patients towards this important but not frequently discussed issue.

Keywords Diet · Nutrition · Chemotherapy · Food–drug interaction · Adverse reactions

Abbreviations

AUC	The area under the concentration versus time curve
Cmax	Maximum concentration
CYP	Cytochrome P450
CYP3A	Cytochrome P3A

R. N. Gacche, *Dietary Research and Cancer*,
https://doi.org/10.1007/978-981-16-6050-4_11

CYP3A4	Cytochrome P3A4
EGCG	Epigallocatechin-3-gallate
GSF	Granulocyte colony-stimulating factors
PD	Pharmacodynamic
PK	Pharmacokinetic

Introduction

The World Health Organization (WHO) has defined complementary medicine as "a broad set of healthcare practices that are not part of a country's own tradition or conventional medicine and are not fully integrated into the dominant healthcare system". In the USA, approximately 36% of cancer patients use CAM: complementary and alternative medicine (Ozben 2015). There is every possibility that the combination of anticancer chemotherapy drugs taken by cancer patients (who are also using CAM) may undergo interactions leading to adverse outcomes. It has been scientifically proved that when dietary supplement or herbal formulations are taken with anticancer drugs, there is always a risk of perturbations on the pharmacokinetic (PK) and pharmacodynamic (PD) properties of the drug (Chan 2013). There are scientific speculations that variety of foods or dietary supplements may act on specific enzymes implicated in the PK of the drug and thereby adversely affect the drug metabolism and transportation. Therefore, it is important to be aware the healthcare professionals about food/diet interactions so that they can help patients to understand judicious usage of CAM in combination with anticancer drugs and thereby can avoid the adverse outcomes. Another reason for inculcating the issue of food–drug interaction is that in the USA and many other counties, dietary supplements are regulated as foods, and not drugs, unless otherwise is claimed for the treatment of specific disease(s). Of note, taking into consideration the public health significance of the issue, the US Food and Drug Administration (fda.gov) has provided "Clinical Drug Interaction Studies-Cytochrome P450 Enzyme- and Transporter-Mediated Drug Interactions Guidance for Industry".

The research finding has proved that the absorption of orally routed anticancer drugs can be altered by metabolic processes in the gastrointestinal tract. The circumstantial literature describes that oral chemotherapy drugs and food interactions are associated with significant impact on the pharmacokinetics of orally administered anticancer agents, which adversely affect the bioavailability, absorption and other pharmacokinetic parameters and possibly limit the benefits of chemotherapy drugs (Segal et al. 2014; Singh and Malhotra 2004). For example, it has been described that the activity of CYP (cytochrome P450) enzymes within the gut wall happens to be a prominent factor in altering the bioavailability profile of orally routed anticancer drugs, which are substrates of CYP3A (Zhang and Benet 2001). A well-investigated example of CYP-mediated changes in bioavailability is the effect of grapefruit juice on bioavailability of anti-inflammatory/immunosuppressive cyclosporine and nifedipine: a calcium channel blocker. Grapefruit juice contains potent inhibitors of intestinal CYP3A4 and hence increases the bioavailability of several drugs leading to symptomatic toxicity (Bailey et al. 1994). Our knowledge and understanding about food–drug interactions are mostly derived from anecdotal experience. More precisely, the dietary interventions may delay, decrease, increase or remain unaffected towards the bioavailability of the drug molecules. For example, both decreased and delayed absorption rates were observed when chlorambucil (an alkylating agent) is administered to leukaemia or lymphoma patients in the fed rather than the fasting condition. The delayed absorption of chlorambucil is associated with the slowed rate of gastric emptying. Moreover, chlorambucil is unstable that readily undergoes hydrolysis in gastric fluid leading to decreased bioavailability (Adair et al. 1986).

Some of the orally given anticancer drugs are prodrugs and require activation in the gastrointestinal tract and/or liver for their anticancer effects, and here, the dietary intervention can influence the bioavailability. For example, a decreased bioavailability by 36 and 63% was observed when estramustine phosphate sodium was administered with food or milk. Therefore, it is imperative that estramustine phosphate sodium should be taken with water 1 h before or 2 h after the meal (Gunnarsson et al. 1990). The food can have differential impacts towards the absorption and bioavailability of orally administrated anticancer drug. For example, the food has demonstrated delayed absorption of series of anticancer drugs such as busulphan, fluorouracil, methotrexate and topotecan, altretamine, capecitabine, chlorambucil, estramustine, gefitinib, melphalan and thioguanine, while the food intervention had increased absorption rate and extent in case of erlotinib and tretinoin. Of note, the food has no effect on absorption (no effect on rate or extent) of etoposide, imatinib, mercaptopurine and temozolomide (Pandey et al. 2017). A recent food–drug interaction network and modelling study revealed that drug–food interactions in cancer and neurological diseases are more prone for demonstrating most negative effects (Jensen et al. 2015).

Overall, the cancer chemotherapy interactions with variety of foods and dietary supplements may have different outcomes. The combination of anticancer drugs and complementary and alternative medicine may undergo interaction and cause adverse reactions.

Certain food constituents and dietary supplements comprising of grapefruit juice, St. John's wort and green tea-derived epigallocatechin gallate can bring alteration in the PK of specific drug types. For example, a combination of grapefruit juice and imatinib resulted in alteration of PK. Scientific literature on interactive outcome of dietary supplement, antioxidants and anticancer drugs is limited; however, there can be several speculated possible adverse reactions (PDQ Integrative, Alternative, and Complementary Therapies Editorial Board 2002).

Clinical Evidences of Food–Drug Adverse Interaction

A vast body of preclinical studies are available in the literature, which has unravelled series of adversities and benefits of combining a specific food/diet or supplement with anticancer drugs (Fasinu and Rapp 2019; Jermini et al. 2019; Wolf et al. 2021). The preclinical data accumulated in the recent past also inspired the research community and pharma sector to design and conduct the clinical trials addressing the issue of food–drug interactions. As a matter of fact, at present (until end of June 2021) there are over 27 clinical trials available at ClinicalTrials.gov investigating the scientific query of food–drug interactions and several reports of completed trials are available (PDQ Integrative, Alternative, and Complementary Therapies Editorial Board 2002).

In a clinical observational study assessing the effect of antioxidant dietary supplement such as multivitamins, selenium, vitamins A, C and E and zinc with anticancer drugs in postmenopausal breast cancer patients, it was observed that the risk of overall mortality was significantly increased along with drastic reduction in recurrence-free survival of cancer patients (Jung et al. 2019). In a similar kind of experimental settings, evaluating the effect of antioxidant supplements (vitamins A, C and E; carotenoids; and coenzyme Q10) with cyclophosphamide, doxorubicin and paclitaxel, the results also demonstrated the increase in mortality and disease recurrence in breast cancer patients (Ambrosone et al. 2020). Alpha-tocopherol, which is one of the eight vitamin E compounds, was tested in a clinical trial for its possible impact on amelioration of side effects of chemotherapy and radiation therapy. In the beginning, the research outcome suggested that alpha-tocopherol may help in reducing toxicity generated by radiation therapy in head and neck cancer patients. However, in two controlled clinical randomized trials of head and neck cancer patients, it was observed that patients who were provided vitamin E supplementation (400 IU/day) have demonstrated a higher risk of tumour relapse and a significant decrease in cancer-free survival (Bairati et al. 2005, 2006).

A case study reported that the consumption of Ginseng with imatinib resulted in severe hepatotoxicity in 26-year-old male cancer patient. The hypothesized mechanism for this adverse outcome may be associated with the inhibition of hepatic CYP3A4, the enzyme primarily implicated in the metabolism of imatinib (Bilgi et al. 2010). There are two well-documented clinical examples of herb–drug interactions adversely affecting the PK of the cancer-preventing drug that has clinical evidence. These two adverse interactions are associated with St. John's wort (SJW and irinotecan and imatinib anticancer drugs). When the patients were administrated with SJW (900 mg/day) and irinotecan (350 mg/m^2), the outcome of the one study reported 42% dropdown in plasma levels of SN-38 (an active metabolite of irinotecan). The researchers hypothesized that the components of SJW extract (pseudohypericin and hyperforin) might have undergone interaction with CYP3A4 and P-gp, which may have resulted in reduction in SN-38 and thus may reduce the efficacy of irinotecan (Mathijssen et al. 2002).

A similar trend of clinical adversities was observed in two studies that were designed to investigate the combinatorial effect of SJW (300 mg 3 times a day) with imatinib (400 mg). The result of one showed that intervention of SJW resulted in 32% decrease in AUC (the area under the concentration versus time curve), a 29% reduction in Cmax and 21% decrease in half-life after 2 weeks of combined treatment with SJW and imatinib (Smith et al. 2004). The outcome of second study observed that inclusion SJW resulted in 43% increase in the clearance of imatinib along with 30% dropdown in AUC. The mechanism of this interaction was also thought to be the interaction of SJW with CYP3A4, which happens to be a key enzyme involved in imatinib metabolism (Frye et al. 2004). Another CYP3A4 substrate that is likely to be altered by SJW exposure is docetaxel. A study conducted with ten cancer patients investigated the PK interactions of combination of SJW (300 mg orally for 14 days) with docetaxel (135 mg IV for 60 min). The results of the study demonstrated statistically significant decrease in mean AUC by 12% along with increased clearance of docetaxel (Goey et al. 2013).

It has been suspected that the combined intake of grapefruit juice and imatinib may increase the plasma levels of imatinib by inhibiting CPY3A4 and may trigger organ toxicity (He et al. 2010).

In a randomized, crossover, pilot study of six participants who were provided with etoposide and grapefruit juice, it was observed that the bioavailability of the etoposide was decreased after coadministration of grapefruit juice. More precisely, the bioavailability of etoposide (50 mg orally) was reduced from 73 to 52% after pretreatment with 100 mL of grapefruit juice. This ultimately resulted in AUC by 26% for etoposide in combination with grapefruit juice as compared with the treatment of etoposide alone (Reif et al. 2002). In a similar kind of experimental settings, the combination of sunitinib with grapefruit juice resulted in increase in the bioavailability of sunitinib (van Erp et al. 2011). Similarly, when nilotinib (400 mg orally) was combined with 240 mL grapefruit juice, the results showed an increase in AUC by 29% and an increase in peak serum concentration (Cmax) by 60% (Yin et al. 2010).

A case study involving the coadministration of EGCG and sunitinib has been described, wherein possible adverse effect of green tea consumption with sunitinib treatment has been reported. A male patient of metastatic renal cell carcinoma taking treatment of sunitinib reported adverse symptoms of eye swelling and hyperemia near the site of a metastatic lesion when the patient was drinking green tea, the improvement in symptom was also observed after he stopped the intake of green tea. The authors speculate that EGCG from green tea may adversely affect the sunitinib's anticancer potential (Ge et al. 2011).

A positive outcome of intake of wheatgrass juice (60 mL orally daily) consumed during chemotherapy treatment using fluorouracil, adriamycin and cytoxan in breast cancer patients has been reported by conducting a pilot randomized clinical trial. The outcome of the trial has demonstrated the significant reduction in myelotoxicity stimulated colony-stimulating factors without any substantial impact on the efficacy of chemotherapy drugs. Such series of clinical reports constantly accumulating in the literature warrants the further clinical research and also alarms the oncologists and

Table 11.1 Profile of food anticancer drug/therapy interactions with commonly used dietary supplement/ingredient/herbal formulation/extract, etc., observed in human clinical trials

Dietary supplement/ ingredient/herbal formulation	Name of the anticancer drug/ therapy	Reported outcome	References
Grapefruit juice	Imatinib	Speculations of increase in plasma levels of imatinib by inhibition of CPY3A4	He et al. (2010)
Grapefruit juice	Etoposide	↓ AUC and ↓ bioavailability of etoposide	Reif et al. (2002)
Grapefruit juice	Sunitinib	↓ Bioavailability of sunitinib	van Erp et al. (2011)
Grapefruit juice	Nilotinib	↓ AUC and C_{max} of nilotinib	Yin et al. (2010)
Green tea infusion	Sunitinib	↓ Drug absorption and bioavail ability of sunitinib and worsened symptoms	Ge et al. (2011)
EGCG	Tamoxifen	↑ Bioavailability of tamoxifen	Shin and Choi (2009)
St. John's wort.	Irinotecan	↑ Activity of CYP3A4 and ↓ AUC of SN38	Mathijssen et al. (2002)
St. John's wort.	Imatinib	↑ Clearance and ↓ AUC of imatinib	Frye et al. (2004)
St. John's wort.	Docetaxel	↑ Clearance and ↓ AUC of docetaxel	Clairet et al. (2019)
Ginseng extract	Imatinib	Incident of hepatotoxicity	Bilgi et al. (2010)
Vitamin E (Dl-alpha-tocopherol)	Radiation therapy	↑ Risk of tumour relapse and ↑ mortality	Bairati et al. (2005); Bairati et al. (2006)
Wheatgrass juice	Fluorouracil, adriamycin and cytoxan)	↓ Myelotoxicity and ↑ activity of GSF	Bar-Sela et al. (2007)

↑: Increase, ↓: decrease, *AUC* the area under the concentration versus time curve

healthcare agencies to take cognizance of food–drug interaction while treating cancer patients. A summary of prominent food anticancer drug interactions reported in final reports of human clinical trials is shown in Table 11.1.

Conclusion

The take-home message from anticancer drug–food interaction discourse is to optimize the management of timings for oral administration of anticancer drugs in relation to timing of taking food like before, after or during a meal. Here comes the role of oncologist and caretakers to guide the cancer patients about dietary intervention so as to improve the chemotherapy outcomes. The food–drug interactions also hold importance in prescribing drug dosage, which plays an important role in optimizing the anticancer activity of the chemotherapy regime and also for minimizing the off-target drug toxicity to normal tissues/organs. The food–drug interactions are affected by a patient's physiological status, genetic makeup and majority of the pharmacokinetic drug interactions are related to drug metabolism and transport. It is important to understand the activity of human enzymes and transporters in drug metabolism and transport. The food–drug interactions are also value additions towards increasing the therapeutic index of anticancer drugs by combinatorial approach as variety of dietary substances target the same drug targets (Scripture and Figg 2006). More importantly, the drug–food interactions if revisited by the oncologists and caretakers at the time of chemotherapy treatment, it also helps in answering the most important questions asked by cancer survivors like which foods should be preferably avoided and prioritized and planning of diet during the oral administration of chemotherapy anticancer drugs. The knowledge of drug–food interactions is also indispensable towards developing a rational personalized anticancer therapeutic approach with reference to the pros and cons of drug–food interactions. This important discourse on clinically significant food/dietary supplement–drug interaction is of immense interest to the public health, especially for the medical practitioners, researchers and patients undergoing cancer chemotherapy.

References

Adair CG, Bridges JM, Desai ZR. Can food affect the bioavailability of chlorambucil in patients with haematological malignancies? Cancer Chemother Pharmacol. 1986;17:99–102.

Ambrosone CB, Zirpoli GR, Hutson AD, et al. Dietary supplement use during chemotherapy and survival outcomes of patients with breast cancer enrolled in a cooperative group clinical trial (SWOG S0221). J Clin Oncol. 2020;38(8):804–14.

Bailey DG, Arnold JM, Spence JD. Grapefruit juice and drugs. How significant is the interaction? Clin Pharmacokinet. 1994;26:91–8.

Bairati I, Meyer F, Gélinas M, et al. A randomized trial of antioxidant vitamins to prevent second primary cancers in head and neck cancer patients. J Natl Cancer Inst. 2005;97(7):481–8.

Bairati I, Meyer F, Jobin E, et al. Antioxidant vitamins supplementation and mortality: a randomized trial in head and neck cancer patients. Int J Cancer. 2006;119(9):2221–4.

Bar-Sela G, Tsalic M, Fried G, Goldberg H. Wheat grass juice may improve hematological toxicity related to chemotherapy in breast cancer patients: a pilot study. Nutr Cancer. 2007;58:43e8.

Bilgi N, Bell K, Ananthakrishnan AN, et al. Imatinib and Panax ginseng: a potential interaction resulting in liver toxicity. Ann Pharmacother. 2010;44(5):926–8.

Chan LN. Drug-nutrient interactions. JPEN J Parenter Enteral Nutr. 2013;37:450–9.

Clairet AL, Boiteux-Jurain M, Curtit E, et al. Interaction between phytotherapy and oral anticancer agents: prospective study and literature review. Med Oncol. 2019;36(5):45.

Fasinu PS, Rapp GK. Herbal interaction with chemotherapeutic drugs—a focus on clinically significant findings. Front Oncol. 2019;9:1356.

Frye RF, Fitzgerald SM, Lagattuta TF, et al. Effect of St John's wort on imatinib mesylate pharmacokinetics. Clin Pharmacol Ther. 2004;76(4):323–9.

Ge J, Tan BX, Chen Y, et al. Interaction of green tea polyphenol epigallocatechin-3-gallate with sunitinib: potential risk of diminished sunitinib bioavailability. J Mol Med (Berl). 2011;89 (6):595–602.

Goey AK, Meijerman I, Rosing H, et al. The effect of Echinacea purpurea on the pharmacokinetics of docetaxel. Br J Clin Pharmacol. 2013;76:467–74.

Gunnarsson PO, Davidsson T, Andersson SB, et al. Impairment of estramustine phosphate absorption by concurrent intake of milk and food. Eur J Clin Pharmacol. 1990;38(2):189–93.

He SM, Yang AK, Li XT, et al. Effects of herbal products on the metabolism and transport of anticancer agents. Expert Opin Drug Metab Toxicol. 2010;6(10):1195–213.

Jensen K, Ni Y, Panagiotou G, Kouskoumvekaki I. Developing a molecular roadmap of drug-food interactions. PLoS Comput Biol. 2015;10(2):e1004048.

Jermini M, Dubois J, Rodondi PY, et al. Complementary medicine use during cancer treatment and potential herb-drug interactions from a cross-sectional study in an academic centre. Sci Rep. 2019;9:5078.

Jung AY, Cai X, Thoene K, et al. Antioxidant supplementation and breast cancer prognosis in postmenopausal women undergoing chemotherapy and radiation therapy. Am J Clin Nutr. 2019;109(1):69–78.

Mathijssen RH, Verweij J, de Bruijn P, et al. Effects of St. John's wort on irinotecan metabolism. J Natl Cancer Inst. 2002;94(16):1247–9.

Ozben T. Antioxidant supplementation on cancer risk and during cancer therapy: an update. Curr Top Med Chem. 2015;15(2):170–8.

Pandey MK, Gupta SC, Nabavizadeh A, Aggarwal BB. Regulation of cell signaling pathways by dietary agents for cancer prevention and treatment. Semin Cancer Biol. 2017;46:158–81.

PDQ Integrative, Alternative, and Complementary Therapies Editorial Board. Cancer therapy interactions with foods and dietary supplements (PDQ®): health professional version. 2021 Jun 10. In: PDQ cancer information summaries [Internet]. Bethesda (MD): National Cancer Institute (US); 2002. https://www.ncbi.nlm.nih.gov/books/NBK563071/.

Reif S, Nicolson MC, Bisset D, et al. Effect of grapefruit juice intake on etoposide bioavailability. Eur J Clin Pharmacol. 2002;58(7):491–4.

Scripture CD, Figg WD. Drug interactions in cancer therapy. Nat Rev Cancer. 2006;6:546–58.

Segal EM, Flood MR, Mancini RS, Whiteman RT, Friedt GA, Kramer AR, Hofstetter MA. Oral chemotherapy food and drug interactions: a comprehensive review of the literature. J Oncol Pract. 2014;10(4):e255–68.

Shin SC, Choi JS. Effects of epigallocatechin gallate on the oral bioavailability and pharmacokinetics of tamoxifen and its main metabolite, 4-hydroxytamoxifen, in rats. Anticancer Drugs. 2009;20(7):584–8.

Singh BN, Malhotra BK. Effects of food on the clinical pharmacokinetics of anticancer agents: underlying mechanisms and implications for oral chemotherapy. Clin Pharmacokinet. 2004;43:1127–56.

Smith P, Bullock JM, Booker BM, et al. The influence of St. John's wort on the pharmacokinetics and protein binding of imatinib mesylate. Pharmacotherapy. 2004;24(11):1508–14.

van Erp NP, Baker SD, Zandvliet AS, et al. Marginal increase of sunitinib exposure by grapefruit juice. Cancer Chemother Pharmacol. 2011;67(3):695–703.

Wolf CPJG, Rachow T, Ernst T, Hochhaus A, Zomorodbakhsch B, Foller S, Rengsberger M, Hartmann M, Hübner J. Interactions in cancer treatment considering cancer therapy, concomitant medications, food, herbal medicine and other supplements. J Cancer Res Clin Oncol. 2021. https://doi.org/10.1007/s00432-021-03625-3.

Yin OQ, Gallagher N, Li A, et al. Effect of grapefruit juice on the pharmacokinetics of nilotinib in healthy participants. J Clin Pharmacol. 2010;50(2):188–94.

Zhang Y, Benet LZ. The gut as a barrier to drug absorption: combined role of cytochrome P450 3A and P-glycoprotein. Clin Pharmacokinet. 2001;40:159–68.

Chapter 12
Functional Foods, Nutraceuticals and Dietary Supplements in Cancer Prevention

Abstract Functional foods comprise foods or food components that have potential to provide health benefits beyond basic requirements of nutrition. It is believed that functional foods are value-added products, which not only provide nutrients but also help to maintain good health and reduce the risk of diseases including cancer. "Nutraceuticals" is a combination of two words "nutrition" and "pharmaceutical", which have demonstrated positive effects on overall health well-being, and also complements for prevention and treatment of diseases including cancer.

Dietary supplements (DSs) are developed to provide micronutrients that otherwise are not consumed in sufficient quantities from the dietary sources. The DSs include micronutrients such as vitamins, minerals, proteins, amino acids or other nutritional botanicals. These products are usually available in the market in the form of tablet, capsule and liquid formulations. The cancer survivors use variety of functional foods, nutraceuticals and DSs with an intension of cancer prevention, maintaining good quality of life and for overall survival. All these three forms of health-promoting agents have demonstrated efficacy in the treatment of different types of cancer. In the present state of the art, the convincing clinical evidence describing the effective role of these complementary and alternative therapeutic approaches is lacking, which warrants the design and conduct of human clinical trials. Nevertheless, the available literature is controversial towards the cancer therapeutic claims. Of note, their nonjudicious consumption has been correlated with variety of cancer-related health adversities.

Keywords Functional foods · Nutraceuticals · Dietary supplements · Cancer prevention

Introduction

The dietary benefits of plant-based foods have been explored in the global market in the name of functional foods, nutraceuticals and dietary supplements (FND). Apart from the consumer catchy and attractive advertisements, there are no universally accepted definitions of these market branded food products. The preclinical and

R. N. Gacche, *Dietary Research and Cancer*,
https://doi.org/10.1007/978-981-16-6050-4_12

clinical research of couple of decades has identified the health-promoting and disease-preventing mechanisms of dietary metabolites, which strongly accelerated the market of FND owing to increasing demands by consumers particularly interested in employing reductionist dietary approaches for health and disease management. In the current state of the art, FND is advertised as foods for the future health and disease management, without taking cognizance of ground realities in "Profit Rich-Evidence Poor" scenario of FND. In other words, the food and pharma industries have taken an advantage of inculcating the significance of FND as an exclusive health promoters and agents to re-equilibrate the unbalanced diet (Fardet 2015; Gul et al. 2015). Asia Pacific, North America, Latin America and Western Europe are the world leaders in the field of FFs. The market of FFs is dynamic and growing rapidly, and therefore, it is difficult to compile the precise estimates of total turnover and the volume of functional foods sold across the world; however, the global market trend shows that the FF market growth rate was over 7.2% (Vicentini et al. 2016). In 2014, the global market value of nutraceuticals was US$ 165.62 billion and it is expected to be reached to US$ 278.96 billion by 2021 with an increased annual growth rate of 7.3% from 2015 to 2021 (Dang 2017). The US$109 billion of the global market was occupied by dietary supplements in 2015, while its estimated market value is US$109 by 2020, which including major shares from Asia Pacific (31%), Europe (30%) and North America (25%) (Binns et al. 2018).

There are also reports that advocate the need-based judicious intake DSs, as it has been observed that the amounts of nutrient intake derived from the foods were found to be far below the range of recommended intake for variety of nutrients such as vitamin E, vitamin D, fibre, potassium and calcium (Zhang et al. 2015). In fact, the research on impact of such shortfall of nutrients on overall health and quality of life in cancer patients has not been well documented. In this regard, there are some suggestions, which state that the low levels of serum marker "vitamin D" have been attributed to the poor survival of cancer patients. Further, a cohort study of 6092 breast cancer patients revealed that the low concentrations of circulating 25-hydroxyvitamin D resulted in high breast cancer-related mortality (Kim and Je 2014). Similarly, another study performed on 1666 breast cancer patients demonstrated that the overall survival rate is low in patients having the lowest levels of tertile of 25-hydroxyvitamin D as compared to those in the highest tertile (Yao et al. 2017). The poor nutrient intake among cancer patients may be associated with the long-lasting impact of the disease and its therapies on dietary intake along with an unmet need. A survey conducted in this regard revealed that nutrition care is severely lacking in the current delivery model of outpatient oncology in the USA (Trujillo et al. 2019). Therefore, advanced research is needed to design effective strategic approaches to integrate diet and nutrition into currently available oncology care to improve the nutritional intake of cancer patients.

The therapeutic implications of the bioactive dietary metabolites constituting FND are already discussed in series of research reports, and the real discourse is about the benefits of these market commodities in prevention of cancer. An overview of dietary bioactive metabolites, fruits, vegetables, whole grains and other substances constituting FNDs clearly demonstrated the preclinical mechanistic anticancer effects against variety of cancers. Moreover, the series of clinical trial outcomes also supports the cancer ameliorative claims of the dietary substances; however, the effects are not qualifying to consider the dietary ingredients or FNDs as therapeutic candidates to get an FDA approval for treating specific cancers. On the contrary, US FDA has evaluated series of anticancer claims of variety of dietary substances constituting FNDs and categorized them into four categories such as (a) "unqualified claims that meet the significant scientific agreement"; (b) "scientific evidence suggests but does not prove"; (c) "limited and not conclusive evidence"; and (d) "very limited and preliminary scientific claim". Majority of the anticancer claims of dietary substances constituting FNDs have been considered as either health claims or conventional food and dietary supplement by FDA regulatory authorities (Hasler et al. 2004; Hasler and Brown 2009). An overview of recent WCRF-AICR dietary guidelines recommended for the management of different types of cancers clearly focuses on the lack of strong convincing therapeutic evidence and dietary guidelines are mostly based on limited suggestive and limited nonconclusive evidence. Nevertheless, with few exceptions majority of the dietary recommendations suggested by different agencies for cancer survivors have not recommended the interventions of dietary supplements or nutraceuticals for cancer prevention, rather emphasized to rely on dietary sources in lieu of dietary supplements.

Role of Functional Foods in Cancer

The functional food market is one of the fastest growing sectors of the worldwide global food market, and it is speculated to excel from about 300 billion US dollars in 2017 to over USD 440 billion in 2022. The USA happens to be the world's largest FFs ($60 billion in 2017), and it is expected to grow by 8% per year (Shahbandeh 2018). USA, Japan, UK, France Netherlands, Germany and Spain are the worldwide leaders in the production of FFs (Vicentini et al. 2016). Functional foods (FFs) are described as the food ingredients that provide positive health benefits beyond the basic requirement of nutrition. FFs share similar appearance with that of traditional foods and are consumed as a part of the normal diet and provide the essential amount of vitamins, proteins, fats and carbohydrates. In summary, FFs represent continuum of ingredients that include natural constituents in conventional, enriched, enhanced and fortified foods (Aghajanpour et al. 2017). According to the American Dietetic Association, the daily consumption of FFs such as vegetables, fruits, plants or seeds, and fortified or enriched food products in varied and effective doses may provide a beneficial effect on human health. The International Food Information Council and International Life Sciences Institute of North America describe that FFs possess proven beneficial effects beyond basic nutrition and are capable of improving health and wellness and/or decrease the risk of disease incidence. Health Canada defines functional foods as "similar in appearance to a conventional food, consumed as part of the usual diet, with demonstrated physiological benefits, and/or to reduce the risk of chronic disease beyond basic nutritional functions" (Hasler and Brown 2009).

Taking a cursory look at these definitions, it is clear that the natural foods such as fruits and vegetables are the simplest form of a functional food. The foods like broccoli, tomatoes or carrots, tea, onion, mushroom, ginger, garlic, apple, ginger, spinach, garlic, green tea, onion cabbage, fennel, soya bean Garlic, green tea and orange come under the category of functional foods owing to their therapeutically rich metabolites like sulforaphane, beta-carotene and lycopene, respectively. Modified foods that have been fortified or enriched with nutrients, phytochemicals, botanicals, etc., also come under the category of FFs. The consumption of FFs is being promoted as one of the cancer prevention strategies (Zeng et al. 2013). It is believed that foods containing functional nutrient are a batter and safer option than usage of dietary supplements, which are more concentrated. Moreover, the complex mixture of nutrients presents in FFs may act synergistically and provide benefits to cancer patients (Serna-Thomé et al. 2018).

Current literature has cited some scientific evidences towards therapeutic benefits of FFs and cancer. For example, a combination of S-ally cysteine from garlic and lycopene from tomatoes has been reported to suppress the progression of chemically induced gastric cancer by modulating apoptosis-related proteins, especially by downregulating the Bcl-2 and upregulating the Bax, Bim and caspases 8 and 3 at considerably lower doses than when these bioactive molecules were given in isolation. Similarly, the lower concentration of the combination of vitamin D3 and genistein has significantly arrested the growth of prostate cancer when compared to individual doses (Aghajanpour et al. 2017). The antiproliferative effects of strawberry tree honey (STH) have been investigated using preclinical cell culture studies on HCT-116 human colon adenocarcinoma and LoVo metastatic cancer cells (Afrin et al. 2019). Further, the extract of ground pistachio kernels (5 mg/mL) inhibited the growth of MCF-7 breast cancer cells by inducing the intracellular ROS generation. The results of the study concluded that the extracts can be used as a "natural adjuvant" in combination with the present chemotherapeutic agents for the treatment of cancers (Reboredo-Rodríguez et al. 2008). There is need to design and conduct RCTs using FFs, so as integrate the FFs as complementary and alternative therapeutic approach.

Nutraceuticals and Cancer Prevention

Nutraceuticals is a hybrid terminology coined by DeFelice in 1989, which describes the nutritional and pharmaceutical aspects of foods. The nutraceuticals are derived from the foods but are sold in various medicinal formulations like capsule, tablet, powder, solution or potion, which have demonstrated physiological health benefits and may provide protection against chronic human ailments including cancer. Series of notable dietary metabolites like β-carotene, lycopene, epigallocatechin-3-gallate, quercetin, curcumin, capsaicin, ellagic acid, gallic acid, pterostilbene, resveratrol, daidzein, genistein, allicin, indole-3-carbinol, isothiocyanates, sulforaphane, caffeine, theophylline and limonene occurring in variety of dietary sources have been studied under preclinical and clinical settings (Clinicaltrials.gov) and branded as cancer-preventive nutraceuticals. These nutraceuticals have also been tested in combination with standard chemotherapy drugs and also demonstrated anticancer effects in late stage of cancer under preclinical model studies. They are also described as possible sensitizers for cancer treatment in radiation therapy (Wargovich et al. 2010; Saldanha and Tollefsbol 2012; Calvaruso et al. 2019). In Canada, these are popularly known as "natural health products". In fact, the boundary between the functional foods and nutraceuticals is not so clear and most of the industries and consumers use it interchangeably. As per the "Dietary Supplements Health and Education Act of 1994" (DSHEA-1994), a dietary supplement is a product taken orally that contains a "dietary ingredient", which supplements the diet. Irrespective of the physical forms of dietary supplements, DSHEA regulation places it in a special category "foods", but not drugs, and also instruct the providers to label it as "dietary supplement" (103rd United States Congress 1994). Contrary to functional foods, dietary supplements and nutraceuticals are made of refined and isolated dietary inclusions, nutrients and/or botanical extracts. Dietary supplements that usually comprises of a multivitamin (mostly C, D, E), minerals, antioxidants, herb or other botanical extracts or metabolites, an amino acid and ω-3 fatty acids and are also sold in various formulations, shapes and sizes such as pills, powders, tablets, capsules and liquids.

Dietary Supplements and Cancer Treatment

A serious discourse about the implications of dietary supplements (DSs) in cancer prevention needs to be revisited in the midst of recent preclinical and clinical findings and the growing market. DSs are reportedly used by over 20–80% of cancer patients after diagnosis. More than 58% of people in the USA consume opine, a dietary supplement for the prevention of cancer. Breast, prostate, colorectal and lung cancers patients have commonly used DSs as a prevention strategy. The role of DSs in the improvement of quality of life and reduced disease progression by overcoming the severe side effects has been documented by the medical practitioners and Internet-motivated self-prescriptions (Paller et al. 2016). Per se, there are several reports describing the role of DSs in cancer prevention; however, the inconsistency in results makes such claims as illusion. The recently published ASC guidelines for diet and physical activity for cancer patients have once again made an appeal of judicious usage of dietary supplements.

The ACS panel has described that there are several products that are frequently marketed under brand name of "Dietary Supplements" but in real sense are not truly "dietary" because several ingredients of such supplements are other than foods sources and are not found in foods consumed by humans. Nevertheless, even they cannot be described as "supplemental" in nature as they are not found to increase the intake of micronutrients, which have been scientifically proved to be indispensable for human health. Moreover, current legitimate regulations and laws do not have any vigilance over such products and do not promise any guarantee that the so-called dietary supplements truly contain the ingredients in the claimed quantities that are mentioned on their pamphlets or the products do not contain any substances that may directly or indirectly can be harmful to human health (Rock et al. 2020).

The outcomes of the prospective clinical trials evaluating the effect of dietary interventions or supplements like curcumin, omega-3 fatty acids, green tea or green tea extracts, melatonin and vitamins C and D in patients with different advanced cancers have been analysed in terms of response rate, disease control rates, time to progression, progression-free survival and overall patient survival. Although there exists a nonsignificant positive correlation with few parameters, the overall impact is much more uncertain and actually is highly controversial. Similar outcome trend for these dietary supplements was observed in phase I, II and III clinical trials when treated in combination with chemodrugs or alone against different cancers (Vernieri et al. 2018). A recent systematic meta-analysis of 49 prospective cancer prevention trials ($n = 287{,}304$) having DS intervention concluded that there was lack of strong conclusive evidence to promote the intervention of DSs as a primary method of preventing deaths from cancer or cardiovascular disease; conversely, under many circumstances the risk of different cancer is even increased (Schwingshackl et al. 2017; World Cancer Research Fund/American Institute for Cancer Research. Continues Update Project Expert Report 2018).

The role of vitamin D supplements has been most extensively elucidated in the prevention of breast, prostate and pancreatic cancer. Series of epidemiological studies have demonstrated that supplemental intake of vitamin D and its higher concentration in blood helps to reduce the risk of colorectal cancer (Woolcott et al. 2010; Ma et al. 2011; Gandini et al. 2011). In contrast, according to Women's Health Initiative RCT the consumption of vitamin D and calcium supplements by healthy women over an average period of 7 years did not result in lowering the incidence of colorectal cancer (Wactawski-Wende et al. 2006). A recently published randomized, placebo-controlled trial enrolling 25,871 participants provided with vitamin D3 at a dose of 2000 IU per day with a median follow-up of 5.3 years concluded that supplementation of vitamin D was not associated with reducing the risk of invasive cancer of any type. However, a post hoc study regarding the rate of death due to cancer showed that the early follow-up and use of DSs revealed possible benefits to the cancer patients (Manson et al. 2019). Further, the cohort study of 5108 cancer patients demonstrated that the high dose (100,000 IU) of vitamin D without calcium over 4 years did not prevent all types of cancer, except skin cancer (Scragg et al. 2018). The RCT designed for investigating the effect of calcium along with vitamin D supplementation on the incidence of cancer in older women, observed that the dose of 25-hydroxyvitamin D (32.8 ng/mL), vitamin D3 and calcium supplementation up to 4 years did not result in reducing the risk of all-type cancer in healthy postmenopausal older women (Lappe et al. 2017).

In the pooled data analysis from two RCTs and one prospective cohort study of 5038 postmenopausal women, it was outlined that the higher levels of 25-hydroxyvitamin D were closely associated with reducing breast cancer risk in a dose-dependent manner and the concentrations of ≥60 ng/mL were found most appropriate for reducing the risk of breast cancer (McDonnell et al. 2018). In summary, considering the current literature linking the role of vitamin D in cancer prevention, it seems difficult to reach a concrete conclusion owing to several controversies associated with vitamin D in relation to cancer prevention. Moreover, several clinical observation studies have shown an inverse relationship between the level of circulating 25(OH)D and the incidence of different types of cancers especially breast, lung, pancreatic, kidney, colorectal, upper gastrointestinal, ovarian and total cancer incidence as well (Marian 2017).

Although substantial volume of evidence advocates the use of calcium supplements in the management of cancer management, however, the evidence claiming the possible benefits from calcium supplements is inconsistent and limited. Therefore, majority of the dietary recommendations does not recommend the use of calcium supplements for cancer prevention. More often, the calcium supplements have been investigated along with vitamin D for assessing the clinical potential in cancer prevention (Wactawski-Wende et al. 2006; Lappe et al. 2017). The studies have been performed on 1200 postmenopausal healthy Nebraska women by delivering the daily dose of 300–600 mg elemental calcium and the same amount in combination with vitamin D supplementation (1000 IU) and a placebo over a period of 4 years. It was observed that the women who were provided with the combination of calcium and vitamin D have a lower (approximately 60%) incidence of cancer when compared to women who administered the placebo. Of note, the lower risk of cancer was also noticed in women who consume only calcium supplements; however, the outcome was not significant on statistical grounds (Lappe et al. 2007). In the ACS's Cancer Prevention Study II Nutrition Cohort enrolling 120,000 men and women, the study outcome revealed that the high intake of dietary calcium and supplements reduced the risk of colorectal cancer. But the effect was not concentration-dependent beyond the daily intake of 1200 mg. The reduction in cancer risk was strongest (31%) for participants who had intake of calcium supplements of 500 mg per day or more (McCullough et al. 2003). Series of studies also demonstrated the negative concerns of calcium supplements and cancer risk. For example, a European Prospective Investigation into Cancer and Nutrition study involving 142,000 men showed that high consumption of calcium or protein from dairy products is responsible for the increased risk of prostate cancer (Allen et al. 2008).

The increased risk of cancer mortality has been reported in a recent prospective cohort study of 30,899 US adults aged 20 years or older after the excess intake of calcium supplements (≥1000 mg/day vs. no use of supplements) (Chen et al. 2019). Thus, the clinical findings of calcium supplements in cancer prevention are inconsistent and nonconclusive. Some of the clinical studies relate the high calcium intake with reduced risk of one or more types of cancer, whereas others link the high calcium intake with increasing the risk of some type of cancers (Marian 2017).

Conclusion

The evolving literature links the functional foods, nutraceuticals and dietary supplements as health-promoting agents and as a complementary and alternative therapeutic modality for the management of cancer. The current OMICS approaches also unravelled the molecular underpinning of these products, which to a greater extent suggest the cancer-preventing role in preclinical studies. The cancer survivors use variety of functional foods, nutraceuticals and DSs with an intension of cancer prevention, maintaining good quality of life and for overall survival. All these three forms of health-promoting agents have demonstrated efficacy in the treatment of different types of cancer. The poor nutrient intake among cancer patients may be associated with the long-lasting impact of the disease and its therapies on dietary intake along with an unmet need. A survey conducted in this regard revealed that nutrition care is severely lacking in the current delivery model of outpatient oncology in the USA (Trujillo et al. 2019). Therefore, further research is required to identify effective strategic approaches to integrate diet and nutrition into clinical oncology care for improving the nutritional intake of cancer patient. In the present state of the art, the convincing clinical evidence describing the effective role of these complementary and alternative therapeutic approaches are lacking, which warrant the design and conduct of human clinical trials. Nevertheless, the available literature is controversial towards the cancer therapeutic claims. Of note, their nonjudicious consumption has been correlated with variety of cancer-related health adversities.

References

103rd United States Congress. The Dietary Supplement Health and Education Act. Public Law. 108 STAT. 1994;4325:103–417.

Afrin S, Giampieri F, Cianciosi D, et al. Strawberry tree honey as a new potential functional food. Part 1: strawberry tree honey reduces colon cancer cell proliferation and colony formation ability, inhibits cell cycle and promotes apoptosis by regulating EGFR and MAPKs signaling pathways. J Funct Foods. 2019;57(7):439–52.

Aghajanpour A, Nazer MR, Obeidavi Z, Akbari M, Ezati P, Kor NM. Functional foods and their role in cancer prevention and health promotion: a comprehensive review. Am J Cancer Res. 2017;7(4):740–69.

Allen NE, Key TJ, Appleby PN, et al. Animal foods, protein, calcium and prostate cancer risk: the European Prospective Investigation into Cancer and Nutrition. Br J Cancer. 2008;98 (9):1574–81.

Binns CW, Lee MK, Lee AH. Problems and prospects: public health regulation of dietary supplements. Annu Rev Public Health. 2018;39:403–20.

Calvaruso M, Pucci G, Musso R, Bravatà V, Cammarata FP, Russo G, Forte GI, Minafra L. Nutraceutical compounds as sensitizers for cancer treatment in radiation therapy. Int J Mol Sci. 2019;20(21). pii: E5267.

Chen F, Du M, Blumberg JB, Ho Chui KK, Ruan M, Rogers G, et al. Association among dietary supplement use, nutrient intake, and mortality among U.S. adults: a cohort study. Ann Intern Med. 2019;170(9):604–13.

Dang R. Nutraceuticals for healthy life. Indian J Pharm Educ Res. 2017;51(3):S148–51.

Fardet A. Complex foods versus functional foods, nutraceuticals and dietary supplements: differential health impact (Part 1). Agro Food Ind Hi Tech. 2015;26(2):20–4.

Gandini S, Boniol M, Haukka J, et al. Meta-analysis of observational studies of serum 25-hydroxyvitamin D levels and colorectal, breast and prostate cancer and colorectal adenoma. Int J Cancer. 2011;128(6):1414–24.

Gul K, Singh AK, Jabeen R. Nutraceuticals and functional foods: the foods for the future world. Crit Rev Food Sci Nutr. 2015;56(16):2617–27.

Hasler CM, Brown AC. American Dietetic Association. Position of the American Dietetic Association: functional foods. J Am Diet Assoc. 2009;109(4):735–46.

Hasler CM, Bloch AS, Thomson CA, Enrione E, Manning C. Position of the American Dietetic Association: functional foods. J Am Diet Assoc. 2004;104(5):814–26.

Kim Y, Je Y. Vitamin D intake, blood 25(OH)D levels, and breast cancer risk or mortality: a meta-analysis. Br J Cancer. 2014;110(11):2772–84.

Lappe JM, Travers-Gustafson D, Davies KM, et al. Vitamin D and calcium supplementation reduces cancer risk: results of a randomized trial. Am J Clin Nutr. 2007;85(6):1586–91.

Lappe J, Watson P, Travers-Gustafson D, Recker R, Garland C, Gorham E, et al. Effect of vitamin D and calcium supplementation on cancer incidence in older women: a randomized clinical trial. JAMA. 2017;317(12):1234–43.

Ma Y, Zhang P, Wang F, et al. Association between vitamin D and risk of colorectal cancer: a systematic review of prospective studies. J Clin Oncol. 2011;29(28):3775–82.

Manson JE, Cook NR, Lee IM, Christen W, Bassuk SS, Mora S, et al. Vitamin d supplements and prevention of cancer and cardiovascular disease. N Engl J Med. 2019;380:33–44.

Marian MJ. Dietary supplements commonly used by cancer survivors: are there any benefits? Nutr Clin Pract. 2017;32(5):607–27.

McCullough ML, Robertson AS, Rodriguez C, et al. Calcium, vitamin D, dairy products, and risk of colorectal cancer in the Cancer Prevention Study II Nutrition Cohort (United States). Cancer Causes Control. 2003;14(1):1–12.

McDonnell SL, Baggerly CA, French CB, Baggerly LL, Garland CF, Gorham ED, et al. Breast cancer risk markedly lower with serum 25-hydroxyvitamin D concentrations >/=60 vs. <20 ng/mL (150 vs. 50 nmol/L): pooled analysis of two randomized trials and a prospective cohort. PLoS One. 2018;13:e0199265.

Paller CJ, Denmeade SR, Carducci MA. Challenges of conducting clinical trials of natural products to combat cancer. Clin Adv Hematol Oncol. 2016;14(6):447–55.

Reboredo-Rodríguez P, González-Barreiro C, Cancho-Grande B, et al. Effect of pistachio kernel extracts in MCF-7 breast cancer cells: inhibition of cell proliferation, induction of ROS production, modulation of glycolysis and of mitochondrial respiration. J Funct Foods. 2008;45(6):155–64.

Rock CL, Thomson C, Gansler T, et al. American Cancer Society guideline for diet and physical activity for cancer prevention. CA Cancer J Clin. 2020;70(4):245–71.

Saldanha SN, Tollefsbol TO. The role of nutraceuticals in chemoprevention and chemotherapy and their clinical outcomes. J Oncol. 2012:192464. https://doi.org/10.1155/2012/192464.

Schwingshackl L, Boeing H, Stelmach-Mardas M, Gottschald M, Dietrich S, et al. Dietary supplements and risk of cause-specific death, cardiovascular disease, and cancer: a systematic review and meta-analysis of primary prevention trials. Adv Nutr. 2017;8:27–39.

Scragg R, Khaw KT, Toop L, Sluyter J, Lawes CMM, Waayer D, et al. Monthly high-dose vitamin D supplementation and cancer risk: a post hoc analysis of the vitamin D assessment randomized clinical trial. JAMA Oncol. 2018;4(11):e182178.

Serna-Thomé G, Castro-Eguiluz D, Fuchs-Tarlovsky V, Sánchez-López M, Delgado-Olivares L, Coronel-Martínez J, et al. Use of functional foods and oral supplements as adjuvants in cancer treatment. Rev Invest Clin. 2018;70(3):136–46.

Shahbandeh M. U.S. functional foods market—statistics & facts. 2018. https://doi.org/https://www.statista.

Trujillo EB, Claghorn K, Dixon SW, Hill EB, Braun A, Lipinski E, Platek ME, Vergo MT, Spees C. Inadequate nutrition coverage in outpatient cancer centers: results of a national survey. J Oncol. 2019;2019:1–8.

Vernieri C, Nichetti F, Raimondi A, Pusceddu S, Platania M, Berrino F, de Braud F. Diet and supplements in cancer prevention and treatment: clinical evidences and future perspectives. Crit Rev Oncol Hematol. 2018;123:57–73.

Vicentini A, Liberatore L, Mastrocola D. Functional foods: trends and development of the global market. Ital J Food Sci. 2016;28:338–51.

Wactawski-Wende J, Kotchen JM, Anderson GL, et al. Calcium plus vitamin D supplementation and the risk of colorectal cancer. N Engl J Med. 2006;354(7):684–96.

Wargovich MJ, Morris J, Brown V, Ellis J, Logothetis B, Weber R. Nutraceutical use in late-stage cancer. Cancer Metastasis Rev. 2010;29(3):503–10.

Woolcott CG, Wilkens LR, Nomura AM, et al. Plasma 25-hydroxyvitamin D levels and the risk of colorectal cancer: the multiethnic cohort study. Cancer Epidemiol Biomark Prev. 2010;19 (1):130–4.

World Cancer Research Fund/American Institute for Cancer Research. Diet, nutrition, physical activity, and cancer: a global perspective. Continues Update Project Expert Report. 2018. dietandcancerreport.org. Accessed 24 Sept 2020.

Yao S, Kwan ML, Ergas IJ, et al. Association of serum level of vitamin D at diagnosis with breast cancer survival: a case-cohort analysis in the pathways study. JAMA Oncol. 2017;3(3):351–7.

Zeng YW, Yang JZ, Pu XY, et al. Strategies of functional food for cancer prevention in human beings. Asian Pac J Cancer Prev. 2013;14(3):1585–92.

Zhang FF, Liu S, John EM, Must A, Demark-Wahnefried W. Diet quality of cancer survivors and noncancer individuals: results from a national survey. Cancer. 2015;121(23):4212–21.

Chapter 13
Do Antioxidants Really Help Cancer Patients? The Puzzle of Benefits and Perils

Abstract The burden of cancer is increasing across the world and accounting for major cause of mortality. Chemotherapy, radiotherapy and surgical procedures are the current approaches for the management of cancer. The preclinical and clinical research suggests that antioxidants are effective in preventing oxidative stress, which is considered as a major culprit for initiation and progression of cancer. Therefore, antioxidants are advised as an adjuvant treatment modality during chemotherapy or radiotherapy treatment of cancer survivors. In the current state of the art, cancer survivors frequently consume the dietary and supplemental antioxidants for the prevention of cancer. Antioxidants are believed to interact with various targets of cancer and arrest cell growth, proliferation, migration, EMT and metastasis and also neutralize the free radicals or reactive oxygen/nitrogen species, which are involved in initiation and progression of cancer. Besides the great strides of advances made in understanding the molecular mechanism of cancer progression, still the concrete mechanisms demonstrating the benefits and adverse effects of antioxidants are poorly understood. However, there is mounting evidence accumulating in the literature, which reports that antioxidant molecules do not demonstrate absolute antitumour or anticancer properties. Nevertheless, few reports also pinpoint the involvement of antioxidants in initiation and progression of cancer. Considering the dynamics of current research outcomes, it seems that antioxidants possess contextual two-faced bioactivity towards cancer prevention and promotion as well. The design and conduct of double-blinded randomized clinical trials can be a gold standard for validating the preclinical anticancer claims of antioxidants and which perhaps will help in integrating the use of antioxidants in the management of cancer. Adequate clinical validation is essential to determine the efficacy of antioxidants or antioxidant supplements in combination with different anticancer agents. The research is evolving in this direction as it can be evidenced by registration (until June 2021) of over 41 clinical trials (ClinicalTrials.gov) out of which 50% are completed.

Keywords Antioxidants · Cancer prevention · Risk and benefits · Clinical trials

R. N. Gacche, *Dietary Research and Cancer*,
https://doi.org/10.1007/978-981-16-6050-4_13

Abbreviations

ABTC	α-Tocopherol/β-Carotene Cancer Prevention Study
AKT	Protein kinase B
BACH1	(BTB domain and CNC homolog 1) is a protein-coding gene
Bcl-2	B-cell lymphoma 2
B-RAF	A proto-oncogene and a serine/threonine-specific protein kinase
CARET	Carotene and Retinol Efficacy Trial
C-myc	Cellular myelocytomatosis (an oncogene)
ECM	Extracellular matrix
EC-SOD	Extracellular EC-SOD
EGCG	Epigallocatechin gallate
EMT	Epithelial-to-mesenchymal transition
GAPDH	Glyceraldehyde-3-phosphate dehydrogenase
GTPase	GTP binding and hydrolyses
Hk2	Hexokinase-2
HOPE-TOO	Heart Outcomes Prevention Evaluation-The Ongoing Outcomes
KRAS	A Kirsten ras oncogene
MnSOD	Manganese superoxide dismutase
MUC-4	Mucin protein-4
NAC	*N*-acetyl cysteine
NF-kB	Nuclear factor κ B
RCT	Randomized controlled trial
RhoA	Ras homolog family member A
ROS	Reactive oxygen species
SELECT	Selenium and Vitamin E Cancer Prevention Trial
SOD	Superoxide dismutase
STAT3	Signal transducer and activator of transcription 3
USPSTF	The US Preventive Services Task Force

Introduction

In a conventional notion, the antioxidants are atoms or molecules that possess ability to neutralize the free radicals, which are generated during the process of metabolism and thereby counteract the process of oxidation. Based on their sources, antioxidants are classified as endogenous antioxidants (enzymatic antioxidant) and exogenous dietary antioxidants such as flavonoids, carotenoids, phenolics, vitamins and minerals (He et al. 2017). The health-promoting antioxidants are naturally available in variety of dietary sources, and their consumption has been associated with variety of health benefits and even for amelioration of human disorders (Yeung et al. 2019). A vast body of literature has accumulated in the recent years, which describe the role of dietary antioxidants in mitigating oxidative stress, which per se is implicated in intuition and progression of several human diseases. Consumption of plant origin nutrients including red wine, organic products, tea, coloured vegetables, nuts, grains and spices happens to be a treasure of dietary antioxidants, which are found effective for maintaining good health, and also acts as complementary therapeutic approach for the prevention of various human diseases including cancer. Different types of minerals such as selenium, copper and zinc and micronutrients vitamins like vitamin E (tocopherol), vitamin A and vitamin C (ascorbic acid) also constitute a class of dietary antioxidants (Goni and Hernandez-Galiot 2019). On the health grounds, the consumption of dietary antioxidant or combined with antioxidant enzymes is strongly recommended for the overall health management. Moreover, several metals such zinc, iron, selenium, manganese and copper act as cofactors of various antioxidant enzymes and dietary nutrients such as α-tocopherol, β-carotene, ascorbic acid and folic acid are known for neutralization or scavenging of reactive oxygen species (Lalkovicova and Danielisova 2016). Besides the availability of effective chemotherapy drugs, which certainly reduce the cancer burden, however, the efficacy of the current drugs is limited due to emerging acquired drug resistance, off-target toxicities and often irreversible side effects. The recent studies also show a hope of antioxidant induced positive response to chemotherapy and amelioration of related side effects. The preclinical studies have reported that some chemotherapy drugs are able to induce oxidative stress and induce apoptosis in normal tissues and also transform cancer cells to a G0 dormant state. It is believed that the antioxidants may suppress oxidative stress and may protect normal cells and escalate the tumour out of dormancy and make it more sensitive towards chemotherapy drugs (Ilghami et al. 2020).

In fact, the antioxidant research, which flourished after 1990s, was basically focused on the role of antioxidants in neutralizing the free radicals. In the later phase of the studies, it was suggested that the free radicals, especially the reactive oxygen species (ROS), play a complex role in different stages of cancer progression. Moreover, the molecular underpinning of ROS induced pathophysiological changes leading to cancer initiation, proliferation, tumour survival, invasion and metastasis is still poorly understood (Hayes et al. 2020). And hence, a strategic plan of consuming antioxidants was proposed to reduce the ROS burden and thereby prevent the cancer incidence and promote antioxidants as a complementary or adjuvant therapeutic approach (Prasad et al. 2017). A repertoire of endogenous noncatalytic small molecules like bilirubin, a-lipoic acid, melatonin, melanin, GSH and uric acid, as well as exogenously consumed vitamin E, vitamin C, b-carotene and plant polyphenols, acts as antioxidants. Moreover, there is also a panel of catalytic antioxidants, which includes copper/zinc superoxide dismutase (SOD), mitochondrial manganese MnSOD (SOD2), extracellular EC-SOD (SOD3), catalase, peroxiredoxins and glutathione peroxidases that work as enzymatic antioxidants (Hayes et al. 2020) (Fig. 13.1).

The understanding of ROS-mediated carcinogenesis is evolving and poorly understood. The chronic inflammation and the crosstalk between the myeloid cell-derived ROS and TNF-alpha-mediated cell signalling are promoting carcinogenesis (Canli et al. 2017). The longer exposure to high ROS leads to DNA damage, while

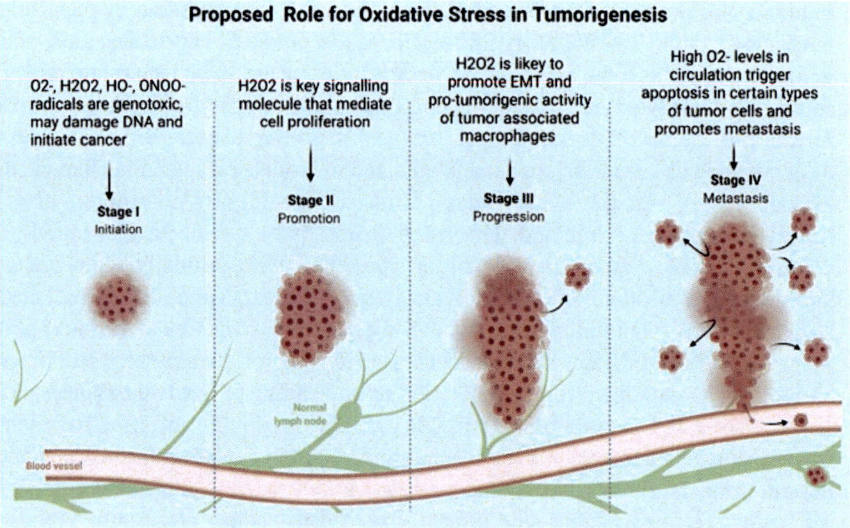

Fig. 13.1 Proposed role for oxidative stress in tumorigenesis. The figure describes the role of oxidative stress in the development of cancer disease from initiation to metastatic stage. The contributions of excess levels of O_2^- H_2O_2, HO^- and $ONOO^-$ at different stages of the disease development are shown at the top panel

the COSMIC mutation induces oxidative DNA damage (Li et al. 2020). The depletion of antioxidant defence mechanisms and the high ROS generation process are responsible for increased cancer risk. Besides there are multiple molecular pathways affected by ROS in general, it has been also reported to play a key role at almost every stage of cancer progression such as initiation, promotion, progression and metastasis (Hayes et al. 2020). While describing the mechanism of action of antioxidants, it has been described that besides neutralizing the ROS and other free radicals they are also reported to inhibit several factors and signalling pathways, which are implicated in cancer progression and metastasis. The variety of factors and signalling pathways inhibited by antioxidants includes STAT-3, MUC-4, NF-kb, C-myc, Bcl2, EMT, Wnt/beta-catenin, Akt, survivin, p53, E-cadherin, B-RAF, KRAS, GTPase and RhoA (Hayes et al. 2020; Dastmalchi et al. 2020).

Series of preclinical and clinical research settings have also demonstrated that the basal oxidative stress of cancer cells is much higher than the normal cells and hence antioxidant system of cancer cells is upregulated and more aggressive in scavenging excess ROS for maintaining cancer-promoting signalling cues and developing resistance to apoptosis (Thyagarajan and Sahu 2018). Conversely, studies also showed that increased ROS levels can also activate anticancer signalling leading to oxidative stress-induced death of cancer cells. The entire discourse also framed a strategy of modifying the ROS generation and antioxidant defences as targets for cancer therapy (Moloney and Cotter 2018). One more important functional attribute of antioxidants demonstrated in several clinical trials is their capabilities to reduce the chemotherapy- and radiotherapy-induced toxicities (Thyagarajan and Sahu 2018).

With advancement of research in the mainstream of antioxidants and cancer prevention, the imbroglio of positive and negative implications of antioxidants in cancer prevention became controversial and more complex. Considering the pre-clinical, clinical research and the systematic reviews accumulated in recent time, at present there exist two schools of thought, the former advocates that the supplemental antioxidant interventions undermine the efficacy of the classical chemo/radiotherapy regimes by adversely affecting their free radical-mediated death of cancer cells, and therefore, the antioxidant supplements should be avoided during the chemo- and radiotherapy episodes. While the other school of thoughts who basically believes in integrative approaches of cancer management, advocates that the supplemental antioxidant interventions during cancer treatment not only helps to alleviate the targeted toxicities and off-target toxicities of the chemo- and radiotherapy approaches, but also enhances the efficacy of cancer therapy (Yasueda et al. 2016; Athreya and Xavier 2017; Thyagarajan and Sahu 2018; Ilghami et al. 2019).

Benefits and Perils of Antioxidant Supplements: Clinical Evidence and Conflicts of Benefits and Risk

The past and current literature has mostly described the implications of supplemental antioxidants like vitamins E (also called α-tocopherol) and C (ascorbic acid), selenium, carotenoids, lycopene, soya products, green tea extract and flavonoids in the management of cancer. In general, it is believed that the proper intake of dietary supplements may help to prevent the overall 35% of cancer cases. More precisely, the fruits and vegetables are rich in antioxidants and act against some types of cancer by different mechanisms. For example, polyphenols, which are known for their free radical scavenging activity demonstrated string antioxidant activity, are reported to prevent the cancer growth in lung, prostate, breast, gastric, tongue, colon and larynx cancers (Singh et al. 2018). Series of preclinical animal and cell culture model studies have reported the beneficial effects of antioxidants (vitamin E, vitamin C, selenium, rutin, quercetin), especially in amelioration of chemotherapy drug-induced toxicity (Gill et al. 2016; Ilghami et al. 2020).

Series of human randomized clinical trials, clinical observational and case–control studies have been conducted, which especially addressed the cancer prevention and toxicity reducing effects of dietary antioxidants (Table 13.1). However, the overall outcome of majority of the trials remains unreliable and without any convincing clinical evidence and advocated further investigations for better understanding. The insufficient period of supplementation has been stated as possible explanation for mainly null outcomes of randomized trials (Conklin 2000).

In a randomized, placebo-controlled, double-blind study, orally supplied vitamin C, selenium and vitamin E to cancer patients treated with cisplatin

Table 13.1 Profile of human randomized clinical trials, clinical observational and case–control studies that have been conducted for addressing the cancer prevention and toxicity reducing effects of dietary antioxidants

Trial ID and type of trial	Number of participants	Chemotherapy drugs/antioxidants	Study outcome	Ref
NCT00342992 (randomized, controlled trials: ABTS trial)	29,133 male smokers of 50–69 years of age in Finland	Vitamin E, beta-carotene, vitamin E + beta-carotene	Increased incidence of lung cancer with b-carotene, along with slight increase in mortality in those who received vitamin E	ATBC Study Group (1994)
NR (clinical study)	79 patients with oral leukoplakia	Beta-carotene, ascorbic acid, alpha-tocopherol	Serum and tissue levels of beta-carotene, ascorbic acid and alpha-tocopherol were significantly increased; but these changes were not correlated with clinical improvement	Kaugars et al. (1994)
NR (a SWOG randomized trial)	153 patients with chronic myelogenous leukaemia	Vitamin A during treatment of busulphan	The patients treated with busulphan plus oral vitamin A had a higher survival rate as compared to a group treated with busulphan alone	Meyskens et al. (1995)
NCT00712647 (randomized, controlled trials: CARET)	18,314 male and female smokers OR male workers exposed to asbestos	Beta-carotene, vitamin A, beta-carotene + vitamin A	Increased incidence of lung cancer and mortality was associated with beta-carotene and vitamin A consumption	Omenn et al. (1996a, b)
NCT00000479 (randomized, controlled trials: WHS)	39,876 female health professionals of age 45	Beta-carotene and Vitamin E	The treatment of antioxidants did not alter the cancer incidence or overall mortality.	Lee et al. (1999, 2005)
NCT00228319 (pilot phase 1/2 clinical trial)	27 ovarian cancer patients	Vitamin C during treatment of paclitaxel, carboplatin	Treatment with vitamin C reduced chemotherapy-related toxicities; however, no significant improvement in overall survival was observed	Ma et al. (2014)

(continued)

Table 13.1 (continued)

Trial ID and type of trial	Number of participants	Chemotherapy drugs/antioxidants	Study outcome	Ref
NR (clinical study)	2 advanced epithelial ovarian cancer patients	Oral vitamin C, vitamin E, beta-carotene, coenzyme Q-10 during treatment of paclitaxel/ carboplatinum	CA-125 was normalized, and the therapeutic efficacy of chemotherapy was improved	Drisko et al. (2003)
NR (clinical study)	31 ovary cancer patients	Selenium supplement during treatment of cisplatin cyclophosphamide	Considerable reduction in hair loss, abdominal pain, flatulence, weakness, malaise and appetite loss was observed	Sieja and Talerczyk (2004)
NR (a randomized, placebo-controlled, double-blind study)	48 patients with solid tumours	Vitamins C and D and selenium supplements during treatment of cisplatin	Protection against organ toxicity, especially nephrotoxicity and ototoxicity, was observed	Weijl et al. (2004)
NCT00272428 (randomized, controlled trials: SU.VI. MAX)	13,017 French females aged 35–60 and males aged 45–60	Vitamin C + vitamin E + beta-carotene + selenium + zinc	Lower cancer incidence and mortality were observed in men, but not in women, and consumption of supplements was associated with increased incidence of skin cancer in women	Hercberg et al. (2004, 2007)
NCT00342654 (randomized, controlled trials: LINXIAN)	29,584 ageing 40- to 69-year-old Linxian residents with no history of cancer or major disease	Retinol + zinc + riboflavin + niacin + vitamin C + molybdenum	Decrease in mortality and lower stomach cancer incidences was observed in group who were taking selenium + vitamin E + beta-carotene	Qiao et al. (2009); Blot et al. (1993)
NCT00006392 (randomized, controlled trials: SELECT)	35,533 males Ageing 50- to 55-year-old w/PSA 4 ng/mL and normal digital rectal exam	Selenium, vitamin E; selenium + vitamin E	No significant difference in prostate cancer incidence was observed, and men supplied with vitamin E alone had a higher incidence of prostate cancer	Lippman et al. (2009); Klein et al. (2011)

(continued)

Table 13.1 (continued)

Trial ID and type of trial	Number of participants	Chemotherapy drugs/antioxidants	Study outcome	Ref
NR (double-blind placebo-controlled trial)	70 patients with chemotherapy-induced peripheral neuropathy	Alpha-lipoic acid during treatment of cisplatin or oxaliplatin	Alpha-lipoic acid administration was not found effective in preventing cisplatin- or oxaliplatin-induced neurotoxicity	Guo et al. (2014)
NR (a pilot study)	43 patients newly diagnosed with solid tumours	Ginger extract supplement during treatment of moderate-to-high emetogenic adjuvant chemotherapy	Ginger extract could increase the blood levels of antioxidant enzyme and decreased oxidative stress blood level. No serious adverse effects were observed after intake of ginger extract as a daily supplement	Danwilai et al. (2017)

NR not registered

chemotherapy, it was observed that supplementation of antioxidants provided protection against organ toxicity, especially against ototoxicity and nephrotoxicity (Weijl et al. 2004). The role of vitamin C in the preventive measure of bleomycin-induced mutagenesis has been elucidated by conducting a clinical study on ten cancer patients. Further, a report showed a prominent reduction in bleomycin-induced chromosomal damage after the consumption of vitamin C doses (100 or 1000 mg) (Pohl and Reidy 1989). Leukoplakia: a white patch lesion on a mucous membrane has been associated with an increased risk of cancer. Proper intake of antioxidant supplements (ascorbic acid, beta-carotene and alpha-tocopherol) in oral leukoplakia patients for 9 months resulted in fast recovery of oral lesions in more than 50% of patients. The tissue levels of ascorbic acid, beta-carotene and alpha-tocopherol were noticed in the antioxidant consumed group (Kaugars et al. 1994).

In an ongoing clinical study, a combination of selenium, vitamin A and lycopene has been provided to prostate cancer patients before the radical prostatectomy to understand the effect of antioxidants on tumour growth using blood biomarkers (unpublished data, Fleshner et al., Clinical Trial No. NCT00844792). The safety studies of ascorbic acid in combination with gemcitabine against pancreatic cancer patients have been well proved in clinical trial phase I. The results showed the rare adverse effects (dry mouth and diarrhoea) of supplementation of ascorbic acid. The survival rate of the patients was around 13 months. Moreover, the combination of vitamin C and gemcitabine was well tolerated by the cancer patients (Welsh et al.

2013). The double-blind placebo-controlled trial has been performed to test the role of alpha-lipoic acid on the severity of peripheral neuropathy symptoms in patients undergoing platinum-based chemotherapy. This study showed that the supplementation of alpha-lipoic acid ALA was not effective in preventing the side effects of chemotherapy (Guo et al. 2014). Further, the study was performed to understand the role of vitamin A in the overall survival of 153 chronic myelogenous leukaemia patients treated with oral pulse busulphan alone and busulphan in combination with vitamin A. This study showed that the patients who consumed vitamin A in a combination of busulphan had a longer duration of clinical survival as compared to patients who took busulphan alone; however, no substantial difference was found at statistical grounds (Meyskens et al. 1995). Such series of clinical trial outcomes have been described in the literature, and few of them are summarized in Table 13.1. It is clear from the outcomes of series of clinical trials that the results are not sufficient to make any definite conclusions. Several clinical trials have been registered (ClinicalTrials.gov), and few are completed, but the outcome of any trials has not been reported.

Few historic investigations that invited the attention of scientific community, oncologist, physicians, caregivers and cancer survivor and other corporate stakeholders are related to the adverse effects of supplemental antioxidants. For example, a study conducted by α-Tocopherol/β-Carotene Cancer Prevention Study (ATBC) group from Finland clinically demonstrated that intake of α-tocopherol (50 mg/per day) and/or β-carotene (20 mg per day) antioxidant supplements over a period of 5–8 years increased the incidence of lung cancer in middle-aged male smokers who had intake of β-carotene supplements. In the subsequent studies, it was also observed that there was no significant impact of either supplement on incidence of pancreatic, urothelial, colorectal, upper aerodigestive tract cancers or renal cancer (The Alpha-Tocopherol Beta Carotene Cancer Prevention Study Group 1994; Rautalahti et al. 1999; Virtamo et al. 2000; Albanes et al. 2000; Wright et al. 2007). In Carotene and Retinol Efficacy Trial (CARET) study, the daily supplementation of 30 mg of β-carotene and 25,000 International Units (IU) of retinol have increased the risk of lung cancer and death in smokers and people exposure to asbestos (Omenn et al. 1996a). Moreover, in later studies the trend of higher risks of lung cancer and all-cause mortality persisted with no effect on risk of prostate cancer (Omenn et al. 1996a; Goodman et al. 2004; Neuhouser et al. 2009). The cohort study performed by a Selenium and Vitamin E Cancer Prevention Trial (SELECT) on 35,533 men over the age of 50 years showed that selenium (200 μg/day) and vitamin E (400 IU/day) supplementation did not reduce the incidence of prostate and other cancers (Klein et al. 2011). In the later phase, the study revealed that more prostate cancer incidence was observed in people who took vitamin E alone (Lippman et al. 2009; Klein et al. 2011). A RCT conducted with 540 head and neck cancer patients treated by radiation therapy was provided with α-tocopherol (400 IU/day) and β-carotene (30 mg/day). The follow-up observations up to median of 6.5 years showed that all-cause mortality was significantly high in the supplement arm as compared to placebo arm (Bairati et al. 2006).

In systematic reviews carried out to understand the effect of supplemental antioxidants on chemotherapeutic efficacy and toxicity, among series of RCTs studied, over 24 studies reported significant decrease in toxicities with concurrent intake of antioxidant supplementation in chemotherapy-treated cancer patients as compared to control patients. In conflict to these findings, nine RCTs concluded no effect of antioxidants on toxicities induced by chemo- or radiotherapy regimes. Of note, one RCT with vitamin A intervention reported increase in toxicity profile (Block et al. 2007, 2008). The current literature also describes that it is difficult to conclude whether antioxidant supplements adversely affect the chemo- or radiotreatment modalities or alleviate the related toxicities (Yasueda et al. 2016; Athreya and Xavier 2017; Thyagarajan and Sahu 2018; Ilghami et al. 2019). The US Preventive Services Task Force (USPSTF) reviewed the outcome of several clinical trials addressing the effect of antioxidant in relation to cancer prevention, toxicity amelioration and side effects. The USPSTF concluded that there were insufficient data for claiming the benefits and risk of antioxidant for cancer prevention. However, the task force mentioned two exceptions, i.e. vitamin E and beta-carotene, wherein they observed that vitamin E has no role in the prevention of cancer, while beta-carotene is responsible for the increased risk of lung cancer in smokers (Moyer and US Preventive Services Task Force 2014).

Besides the above described RCTs, cohort, case–control and observational studies focusing on the adversities associated with supplemental antioxidants, there are many similar settings, which suggest no effect of supplemental antioxidants on cancer incidence and prevention. For example, in Physicians' Health Study I (PHS I, $n = 22{,}071$), the supplementation of β-carotene (50 mg after every other day) followed up to 13 years revealed no effect on relative risk of prostate, lung and colon cancer either it affected cancer-related mortality or all-cause mortality in smokers or nonsmoker participants (Hennekens et al. 1996; Cook et al. 2000). There were no significant changes in the PSA levels observed after the given dose of lycopene (15 mg/day) and green tea catechins (EGCG, 600 mg/day) in a randomized placebo-controlled trial phase II, which was carried out on 469 high-risk prostate cancer patients (Lane et al. 2018). Heart Outcomes Prevention Evaluation-The Ongoing Outcomes (HOPE–TOO) study showed that the intake of α-tocopherol (400 IU/day) over a period of 7 years did not have any effect on cancer occurrence or related mortality (Lonn et al. 2005). The Women's Health Study (WHS) showed that the intake of β-carotene, vitamin E and aspirin for 2 years by 45 years and the older aged women have no harmful effect and benefit by intake of β-carotene or vitamin E supplementation (Lee et al. 2005).

In the current state-of-the-art literature, the similar trend of inconsistency, conflicting and controversial outcome of antioxidant supplements, was observed in several preclinical and clinical investigations. A recent preclinical study published in "Cell" focuses on the promises and perils of antioxidant supplementation in mice models. This study demonstrated that antioxidants stimulate the KRAS-mediated lung cancer metastasis by stabilizing the transcription factor BACH1, which otherwise activates transcription of Hk2 and Gapdh genes and triggers the glycolysis-mediated metastasis. Further, it was suggested that the targeting of BACH1 or its glycolytic protein targets might prevent antioxidant-mediated metastasis (Wiel et al. 2019). This work has got a public perception that "Antioxidants may encourage the spread of lung cancer rather than prevent it" (https://www.sciencenews.org/article/antioxidants-lung-cancer-spread-prevent). Furthermore, the mouse model study showed that the *N*-acetyl cysteine (NAC) was responsible for the increased lymph node metastases in mouse models having malignant melanoma.

Similarly, NAC and the Trolox (a soluble vitamin E analogue) remarkably increased the migration and invasive potential of human malignant melanoma cells. The study results are in conflict with the previous claims of antioxidant effects on metastasis processes such as migration and invasion (Gal et al. 2015). Apart from small-molecule antioxidants, the antioxidant enzymes such as catalase and superoxide dismutase are also reported to facilitate the survival of breast cancer cells after extracellular matrix detachment (ECM). It was also observed that antioxidant enzyme knockdown cancer cells were compromised in their abilities to generate tumours in mice. The authors suggest that inhibiting the antioxidant enzyme activity can be tailored as a druggable target to make the cancer cells more vulnerable towards cell death, which otherwise may survive ECM detachment (Davison et al. 2013). Recent systematic reviews and meta-analysis suggested possible positive implications of green tea in prostate cancer amelioration, but the circumstantial evidence is conflicting for overall prostate cancer incidence (Guo et al. 2017). A third update of the Cochrane review "Selenium for preventing cancer" analysed 83 RCTs involving 27,232 participants allocated to selenium supplements or placebo. The authors concluded that no positive effect of selenium supplements has been observed in reducing the cancer risk. Additionally, conflicting outcomes have been reported concerning supplementation of selenium and higher incidence of prostate cancer and type 2 diabetes (Vinceti et al. 2018).

Apart from the negative reports of antioxidant supplements, the contemporary literature also describes the positive outcomes of the antioxidant applications in cancer prevention. In a systematic prospective cohort setting involving 56,048 human volunteers of the Danish Diet and the study duration of 23 years, the positive effects of antioxidant flavonoids (1000 mg/day) were found to be associated with involved in reduced cancer-related mortality (Bondonno et al. 2019). The study has been performed to investigate the relationship between antioxidant vitamins (mainly α-carotene, β-carotene, and vitamins E and C) and the risk of invasive cervical cancer in 458 Chinese women. The results indicated that antioxidant vitamins are useful to reduce the risk of cervical cancer invasiveness and the effect was found more prominent in passive smokers (Guo et al. 2015).

A meta-analysis of 69 research investigations revealed that the risk of cancer was reduced after exposure to high selenium. The outcome of linear and nonlinear dose–response analysis indicated that the high serum or plasma selenium and toenail selenium levels were associated with cancer prevention potential of selenium (Cai et al. 2016). Several studies indicated that the intake of lycopene and circulating lycopene levels was observed to be responsible for the reduced risk of prostate cancer (Wang et al. 2015; Xu et al. 2016). In recently published 69 prospective studies involving dietary intake and blood concentrations of vitamins C and E and carotenoids concluded that the high dietary intake and/or blood concentrations of vitamin C, α-tocopherol and carotenoids were closely linked to reduced cancer risk and all-cause mortality. The authors encourage more consumption of fruit and vegetable but discourage the use of antioxidant supplements for the management of cancer (Aune et al. 2018). A primary prevention trial enrolling 29,584 adults of both sexes was evaluated for the effect of antioxidant supplements like vitamin E (30 mg), β-carotene (15 mg) and selenium (50 μg) over a period of 10 years. Interestingly, the study outcome found reduced mortality due to stomach cancer (Qiao et al. 2009). Of note, the supplementation of fish oil (omega-3 fatty acid) significantly increased the enzymatic and non-enzymatic status of plasma antioxidant status in the breast cancer patients (Mansara et al. 2015).

Conclusion

A cursory look at the circumstantial literature cited above clearly focuses on the conflicting outcomes about the inclusion of antioxidant supplements as cancer prevention modality. In general, the negative aspects of antioxidant intake show strong evidence of increasing the cancer incidence in several RCTs, while the evidence of benefits of supplemental antioxidants in cancer prevention is comparatively poor, uncertain and nonconclusive. Once again, the ASC guidelines for "diet and physical activity" published in 2020 clearly advocate the importance of consuming antioxidants through dietary food resources rather than intake of marketed supplements. The expert panel also clarified the people who consume variety of fruits and vegetables rich in antioxidant sources and may reduce the risk of some types of cancer; however, it cannot be concluded that the benefits of fruits and vegetables are primarily associated with their antioxidant contents, but it can be from other bioactive food ingredients. The panel reiterated that the clinical evidence is not in favour of antioxidant supplements towards reducing the risk of cancer; on the contrary, some investigations have demonstrated the increased risk of cancer that is frequently exposed to antioxidant supplements (Rock et al. 2020). Perhaps, the inconsistencies and conflicting outcomes might be associated with multiple variables, including but may not be limited to endogenous antioxidant levels in cancer patient, type of antioxidant supplement, dose of antioxidant, route of administration, bioavailability profile, timing of antioxidant intervention, polymorphism in the antioxidant enzymes, dose of chemotherapy, mechanism of action of chemotherapeutic drug and pathophysiology of cancer progression. Filling the gaps carefully with defined RTCs incorporating the above factors might rationalize the judicious interventions of antioxidant supplements towards cancer prevention. But in the current state of art, the notion of "antioxidant supplements for cancer prevention" is controversial, conflicting, perhaps alarming and warrants rational usage. Of note, there is also a hue and cry about the nonjudicious clinical prescriptions of supplemental antioxidants without referring the adverse or ill effects of higher doses of antioxidant supplements (Kalra 2019). PS: It is advisable to cancer patients and healthcare providers to utilize an evidence-based antioxidant supplement for cancer patients under the supervision of professional expert.

References

Albanes D, Malila N, Taylor PR, et al. Effects of supplemental alpha-tocopherol and beta-carotene on colorectal cancer. Cancer Causes Control. 2000;11(3):197–205.

ATBC. The effect of vitamin E and b carotene on the incidence of lung cancer and other cancers in male smokers. The Alpha-Tocopherol Beta Carotene Cancer Prevention Study Group. N Engl J Med. 1994;330:1029–103.

Athreya K, Xavier MF. Antioxidants in the treatment of cancer. Nutr Cancer. 2017;69 (8):1099–104.

Aune D, Keum NN, Giovannucci E, Fadnes LT, Boffetta P, Greenwood DC, et al. Dietary intake and blood concentrations of antioxidants and the risk of cardiovascular disease, total cancer, and all-cause mortality: a systematic review and dose-response meta-analysis of prospective studies. Am J Clin Nutr. 2018;108:1069–91.

Bairati I, Meyer F, Jobin E, Gélinas M, Fortin A, et al. Antioxidant vitamins supplementation and mortality: a randomized trial in head and neck cancer patients. Int J Cancer. 2006;119 (9):2221–4.

Block KI, Koch AC, Mead MN, Tothy PK, Newman RA, Gyllenhaal C. Impact of antioxidant supplementation on chemotherapeutic efficacy: a systematic review of the evidence from randomized controlled trials. Cancer Treat Rev. 2007;33:407–18.

Block KI, Koch AC, Mead MN, Tothy PK, Newman RA, Gyllenhaal C. Impact of antioxidant supplementation on chemotherapeutic toxicity: a systematic review of the evidence from randomized controlled trials. Int J Cancer. 2008;123:1227–39.

Blot WJ, Li JY, Taylor PR, et al. Nutrition intervention trials in Linxian, China: supplementation with specific vitamin/mineral combinations, cancer incidence, and disease-specific mortality in the general population. J Natl Cancer Inst. 1993;85:1483–92.

Bondonno NP, Dalgaard F, Kyro C, Murray K, Bondonno CP, et al. Flavonoid intake is associated with lower mortality in the Danish Diet Cancer and Health Cohort. Nat Commun. 2019;10 (1):3651.

Cai X, Wang C, Yu W, Fan W, Wang S, Shen N, et al. Selenium exposure and cancer risk: an updated meta-analysis and metaregression. Sci Rep. 2016;6:19213.

Canli O, Nicolas AM, Gupta J, et al. Myeloid cell-derived reactive oxygen species induce epithelial mutagenesis. Cancer Cell. 2017;32:869–883.e5.

Conklin KA. Dietary antioxidants during cancer chemotherapy: impact on chemotherapeutic effectiveness and development of side effects. Nutr Cancer. 2000;37:1–18.

Cook NR, Le IM, Manson JE, Buring JE, Hennekens CH. Effects of beta-carotene supplementation on cancer incidence by baseline characteristics in the Physicians' Health Study (United States). Cancer Causes Control. 2000;11:617–26.

Danwilai K, Konmun J, Sripanidkulchai B-O, et al. Antioxidant activity of ginger extract as a daily supplement in cancer patients receiving adjuvant chemotherapy: a pilot study. Cancer Manag Res. 2017;9:11.

Dastmalchi N, Baradaran B, Latifi-Navid S, et al. Antioxidants with two faces toward cancer. Life Sci. 2020;258:118186.

Davison CA, Durbin SM, Thau MR, Zellmer VR, Chapman SE, Diener J, et al. Antioxidant enzymes mediate survival of breast cancer cells deprived of extracellular matrix. Cancer Res. 2013;73(12):3704–15.

Drisko JA, Chapman J, Hunter VJ. The use of antioxidants with first-line chemotherapy in two cases of ovarian cancer. J Am Coll Nutr. 2003;22:118–23.

Gill JG, Piskounova E, Morrison SJ. Cancer, oxidative stress, and metastasis. Cold Spring Harb Symp Quant Biol. 2016;81:163–75.

Goni I, Hernandez-Galiot A. Intake of nutrient and non-nutrient dietary antioxidants. Contribution of macromolecular antioxidant polyphenols in an elderly Mediterranean population. Nutrients. 2019;11(9):2165.

Goodman GE, Thornquist MD, Balmes J, et al. The Beta-Carotene and Retinol Efficacy Trial: incidence of lung cancer and cardiovascular disease mortality during 6-year follow-up after stopping beta-carotene and retinol supplements. J Natl Cancer Inst. 2004;96(23):1743–50.

Guo Y, Jones D, Palmer JL, et al. Oral alpha-lipoic acid to prevent chemotherapy induced peripheral neuropathy: a randomized, double-blind, placebo-controlled trial. Support Care Cancer. 2014;22:1223–31.

Guo L, et al. Associations between antioxidant vitamins and the risk of invasive cervical cancer in Chinese women: a case-control study. Sci Rep. 2015;5:13607.

Guo Y, Zhi F, Chen P, Zhao K, Xiang H, Mao Q, et al. Green tea and the risk of prostate cancer: a systematic review and meta-analysis. Medicine. 2017;96:e6426.

Hayes JD, Dinkova-Kostova AT, Tew KD. Oxidative stress in cancer. Cancer Cell. 2020;38 (2):167–97.

He L, He T, Farrar S, et al. Antioxidants maintain cellular redox homeostasis by elimination of reactive oxygen species. Cell Physiol Biochem. 2017;44(2):32–553.

Hennekens CH, Buring JE, Manson JE, Stampfer M, Rosner B, Cook NR, et al. Lack of effect of long-term supplementation with beta carotene on the incidence of malignant neoplasms and cardiovascular disease. N Engl J Med. 1996;334:1145–9.

Hercberg S, Galan P, Preziosi P, et al. The SU.VI.MAX Study: a randomized, placebo-controlled trial of the health effects of antioxidant vitamins and minerals. Arch Intern Med. 2004;164:2335–42.

Hercberg S, Ezzedine K, Guinot C, et al. Antioxidant supplementation increases the risk of skin cancers in women but not in men. J Nutr. 2007;137:2098–105.

Ilghami R, Barzegari A, Mashayekhi MR, Letourneur D, Crepin M, Pavon-Djavid G. The conundrum of dietary antioxidants in cancer chemotherapy. Nutr Rev. 2019. pii: nuz027. https://doi.org/10.1093/nutrit/nuz027.

Ilghami R, Barzegari A, Mashayekhi MR, et al. The conundrum of dietary antioxidants in cancer chemotherapy. Nutr Rev. 2020;78(1):65–76.

Kalra BS. Supplemental antioxidants: a hype in disease prevention. J Assoc Physicians India. 2019;67(8):66–8.

Kaugars GE, Silverman S Jr, Lovas JG, et al. A clinical trial of antioxidant supplements in the treatment of oral leukoplakia. Oral Surg Oral Med Oral Pathol. 1994;78:462–8.

Klein EA, Thompson IM Jr, Tangen CM, et al. Vitamin E and the risk of prostate cancer: the Selenium and Vitamin E Cancer Prevention Trial (SELECT). JAMA. 2011;306:1549–56.

Lalkovicova M, Danielisova V. Neuroprotection and antioxidants. Neural Regen Res. 2016;11 (6):865.

Lane JA, Er V, Avery KNL, Horwood J, Cantwell M, Caro GP, et al. ProDiet: a phase II randomized placebo-controlled trial of green tea catechins and lycopene in men at increased risk of prostate cancer. Cancer Prev Res. 2018;11(11):687–96.

Lee IM, Cook NR, Manson JE, et al. Beta-Carotene supplementation and incidence of cancer and cardiovascular disease: the Women's Health Study. J Natl Cancer Inst. 1999;91:2102–6.

Lee IM, Cook NR, Gaziano JM, et al. Vitamin E in the primary prevention of cardiovascular disease and cancer: the Women's Health Study: a randomized controlled trial. JAMA. 2005;294:56–65.

Li R, Halliwill KD, Adams CJ, et al. Mutational signatures in tumours induced by high and low energy radiation in Trp53 deficient mice. Nat Commun. 2020;11:394.

Lippman SM, Klein EA, Goodman PJ, et al. Effect of selenium and vitamin E on risk of prostate cancer and other cancers: the Selenium and Vitamin E Cancer Prevention Trial (SELECT). JAMA. 2009;301(1):39–51.

Lonn E, Bosch J, Yusuf S, et al. Effects of long-term vitamin E supplementation on cardiovascular events and cancer: a randomized controlled trial. JAMA. 2005;293(11):1338–47.

Ma Y, Chapman J, Levine M, et al. High-dose parenteral ascorbate enhanced chemosensitivity of ovarian cancer and reduced toxicity of chemotherapy. Sci Transl Med. 2014;6:222ra18.

Mansara P, Ketkar M, Deshpande R, et al. Improved antioxidant status by omega 3 fatty acid supplementation in breast cancer patients undergoing chemotherapy: a case series. J Med Case Rep. 2015;9:148.

Meyskens FL, Kopecky KJ, Appelbaum FR, et al. Effects of vitamin A on survival in patients with chronic myelogenous leukemia: a SWOG randomized trial. Leuk Res. 1995;19:605–12.

Moloney JN, Cotter TG. ROS signalling in the biology of cancer. Semin Cell Dev Biol. 2018;80:50–64.

Moyer VA, US Preventive Services Task Force. Vitamin, mineral, and multivitamin supplements for the primary prevention of cardiovascular disease and cancer: U.S. Preventive Services Task Force recommendation statement. Ann Intern Med. 2014;160:558–64.

Neuhouser ML, Barnett MJ, Kristal AR, et al. Dietary supplement use and prostate cancer risk in the Carotene and Retinol Efficacy Trial. Cancer Epidemiol Biomark Prev. 2009;18(8):2202–6.

Omenn GS, Goodman GE, Thornquist MD, et al. Effects of a combination of beta carotene and vitamin A on lung cancer and cardiovascular disease. N Engl J Med. 1996a;334(18):1150–5.

Omenn GS, Goodman GE, Thornquist MD, et al. Risk factors for lung cancer and for intervention effects in CARET, the b-Carotene and Retinol Efficacy Trial. J Natl Cancer Inst. 1996b;88:1550–9.

Pohl H, Reidy JA. Vitamin C intake influences the bleomycin-induced chromosome damage assay: implications for detection of cancer susceptibility and chromosome breakage syndromes. Mutat Res. 1989;224:247–52.

Prasad S, Gupta SC, Tyagi AK. Reactive oxygen species (ROS) and cancer: role of antioxidative nutraceuticals. Cancer Lett. 2017;387:95–105.

Qiao YL, Dawsey SM, Kamangar F, et al. Total and cancer mortality after supplementation with vitamins and minerals: follow-up of the Linxian General Population Nutrition Intervention Trial. J Natl Cancer Inst. 2009;101:507–18.

Rautalahti MT, Virtamo JR, Taylor PR, et al. The effects of supplementation with alpha-tocopherol and beta-carotene on the incidence and mortality of carcinoma of the pancreas in a randomized, controlled trial. Cancer. 1999;86(1):37–42.

Rock CL, Thomson C, Gansler T, et al. American Cancer Society guideline for diet and physical activity for cancer prevention. CA Cancer J Clin. 2020;70(4):245–71.

Sieja K, Talerczyk M. Selenium as an element in the treatment of ovarian cancer in women receiving chemotherapy. Gynecol Oncol. 2004;93:320–7.

Singh K, Bhori M, Kasu YA, et al. Antioxidants as precision weapons in war against cancer chemotherapy induced toxicity—exploring the armoury of obscurity. Saudi Pharm J. 2018;26 (2):177–90.

The Alpha-Tocopherol Beta Carotene Cancer Prevention Study Group. The effect of vitamin E and beta carotene on the incidence of lung cancer and other cancers in male smokers. N Engl J Med. 1994;330:1029–35.

Thyagarajan A, Sahu RP. Potential Contributions of Antioxidants to Cancer Therapy: immunomodulation and radiosensitization. Integr Cancer Ther. 2018;17(2):210–6.

Vinceti M, Filippini T, Del Giovane C, Dennert G, Zwahlen M, Brinkman M, et al. Selenium for preventing cancer. Cochrane Database Syst Rev. 2018;1:CD005195. https://doi.org/10.1002/14651858.CD005195.pub4.

Virtamo J, Edwards BK, Virtanen M, et al. Effects of supplemental alpha-tocopherol and beta-carotene on urinary tract cancer: incidence and mortality in a controlled trial (Finland). Cancer Causes Control. 2000;11(10):933–9.

Wang Y, Cui R, Xiao Y, Fang J, Xu Q. Effect of carotene and lycopene on the risk of prostate cancer: a systematic review and dose-response meta-analysis of observational studies. PLoS One. 2015;10:e0137427.

Weijl NI, Elsendoorn TJ, Lentjes EG, Hopman GD, et al. Supplementation with antioxidant micronutrients and chemotherapy-induced toxicity in cancer patients treated with cisplatin-based chemotherapy: a randomised, double-blind, placebo-controlled study. Eur J Cancer. 2004;40(11):1713–23.

Welsh J, Wagner B, Van't Erve T, et al. Pharmacological ascorbate with gemcitabine for the control of metastatic and node-positive pancreatic cancer (PACMAN): results from a phase I clinical trial. Cancer Chemother Pharmacol. 2013;71:765–75.

Wiel C, Le Gal K, Ibrahim MX, Jahangir CA, Kashif M, Yao H, et al. BACH1 stabilization by antioxidants stimulates lung cancer metastasis. Cell. 2019;178(2):330–45.

Wright ME, Virtamo J, Hartman AM, et al. Effects of alpha-tocopherol and beta-carotene supplementation on upper aerodigestive tract cancers in a large, randomized controlled trial. Cancer. 2007;109(5):891–8.

Xu X, Li J, Wang X, Wang S, Meng S, Zhu Y, et al. Tomato consumption and prostate cancer risk: a systematic review and meta-analysis. Sci Rep. 2016;6:37091.

Yasueda A, Urushima H, Ito T. Efficacy and interaction of antioxidant supplements as adjuvant therapy in cancer treatment: a systematic review. Integr Cancer Ther. 2016;15(1):17–39.

Yeung AWK, Tzvetkov N, et al. Antioxidants: scientific literature landscape analysis. Oxidative Med Cell Longev. 2019;2019:2–13.

Chapter 14
Market of Dietary Supplements: Analysis of Health Benefits and Risk in Cancer

Abstract In the current state of the art, there is widespread usage of dietary supplements (DSs) by cancer patients both during the cancer treatment and during the drug holidays as well. However, there is lack of empirical data in relation to their safety or efficacy. The available research data show that there should be judicious usage of DSs by cancer patients, which otherwise may aggravate the progression of cancer. Besides these ground clinical realities about the benefits of DSs in cancer prevention, the market of DSs is growing in logarithmic fashion. US people spent over $36 billion in 2014 on dietary supplements. In fact, the consumer victimizing advertisements of DSs are easily accessed by cancer survivors through Internet, social media and television and inspire them to use it as an adjunct or traditional therapy. These self-prescriptions of DSs by cancer patients are mostly based on anecdotal evidences. More often owing to the drastic change in the psychology and mindset of cancer patients, they do not inform the inclusions of DSs to oncologist or their primary caretakers. There is need to rationalize the intake of DSs and quantify the does by assessing the analytical profile of need and deficiency in individual cancer type/patient so as to avoid the DS-related adversities and minimize the financial burden.

Keywords Dietary supplements · Guidelines · Health benefits · Risk in cancer · Market

Abbreviations

ABTS	Alpha-Tocopherol, Beta-Carotene Cancer Prevention
ACS	American Cancer Society
AICR	American Institute for Cancer Research
ASN	American Society for Nutrition
CARET	Beta-Carotene and Retinol Efficacy Trial
CCS	Canadian Cancer Society
CRUK	Cancer Research UK
CSC	Cancer Support Community

R. N. Gacche, *Dietary Research and Cancer*,
https://doi.org/10.1007/978-981-16-6050-4_14

DS	Dietary supplements
DSHEA	Dietary Supplement Health and Education Act
ESPEN	The European Society for Clinical Nutrition and Metabolism
NCCN	National Comprehensive Cancer Network
NCI	National Cancer Institute
NHANES	National Health and Nutrition Examination Survey
SELECT	The Selenium and Vitamin E Cancer Prevention Trial
WCRF	World Cancer Research Fund
WHO	World Health Organization
WHO: ECAC	European Code Against Cancer

Introduction

The global dietary supplement (DS) market was estimated to be around USD 101.38 billion in 2018, which was speculated to double by 2020 (approx. USD 220.3 billion), with further trend of growing market (Hys 2020). DSs mostly available in the current market consist of minerals, vitamins, amino acids or other so-called health substance having some or other physiological effects. There is logarithmic growth of DS market all over the world and is steadily increasing since last decades. Approximately, over 50–75% of populations consume DSs routinely, and approximately half of them consume on regular basis (Cowan et al. 2018; Sekhri and Kaur 2014; Barnes et al. 2016; Czerwiński and Liebers 2019). Consumers of DSs have a broad range of realistic and rebuttal reasons, and it varies with their sex, age, physical activity, health status, etc., but overall the consumers of DSs relate its usage with health and wellness, for prevention of illness and also for amelioration of the dietary deficiencies (Dickinson et al. 2014; Barnes et al. 2016; Hys 2020). Interestingly, sizable number of consumers use DSs devoid of any clinical signs or symptoms of deficiency, perhaps such reluctant usage of DSs may create health risk in long-term situation, and it is essential to counsel such amateur consumers that DSs cannot be a substitute for healthy and balanced diet, but it can be only consumed for the correction of deficiency or for the management-specific deficiency-related human disorders under the supervision of medical professionals (Zhang et al. 2020).

There are also reports, which advocate the need-based judicious intake DSs, as it has been observed that the quantity of nutrient intake derived from the foods was found to be far below the range of recommended intake for variety of nutrients such as vitamin E, vitamin D, fibre, potassium and calcium (Zhang et al. 2015). In fact, the research on impact of such shortfall of nutrients on overall health and quality of life in cancer patients has not been well documented. In this regard, there are some suggestions, which state that the low levels of vitamin D have been considered as serum marker, which is associated with impaired survival. A meta-analysis of 6 prospective cohort studies involving 6092 patients of breast cancer revealed that low levels of concentrations of 25-hydroxyvitamin D in blood were significantly affiliated with increased risk of breast cancer-related mortality (Kim and Je 2014). These findings have been further validated and confirmed by a study involving 1666 breast cancer patients who were reported to have worst overall survival among patients in the lowest tertile of 25-hydroxyvitamin D concentrations as compared to those in the highest tertile (Yao et al. 2017). The poor nutrient intake among cancer patients may be associated with the long-term impact of the disease and its therapies, along with an unfulfilled dietary requirement. A survey conducted in this regard disclosed that the nutrition care is highly inadequate in the current delivery model of outpatient oncology in the USA (Trujillo et al. 2019). Therefore, further research is mandatory to determine effective strategic approaches to amalgamate diet and nutrition into clinical oncology care for improving the overall nutritional intake by cancer patients.

The emotional status of cancer and its assumption as a capital death compels and inspires the cancer patients to adopt multiple therapeutic approaches (both scientific and traditional) for the cure of the disease. More precisely, the most commonly used DSs consumed by cancer patients for the prevention, risk minimization, avoiding relapse and mortality range from supplement of vitamins/minerals (multivitamins/ minerals, vitamin C, D, E, etc.), botanical supplements (astragalus, Silybum marianum, turmeric, resveratrol, ginger, ginseng, lycopene, etc.) and others (-β-carotene ω-3, fatty acids, glutamine and melatonin). The cancer patients often initiate dietary modifications (including dietary supplements) after the cancer diagnosis and during the treatment period (Satia et al. 2009). In general, it has been reported that over 15–50% of cancer patients start the intake of new supplement after diagnosis with a hope of better prognosis and quality of life (Velicer and Ulrich 2008; Patterson et al. 2003). On one hand, the nutrition guidelines recommended by various agencies recommend that cancer survivors can achieve adequate sources of dietary supplements through daily intake of healthy balanced diet, and per se, there is no need to rely on supplements unless otherwise is it needs (Rock et al. 2012). However, on the other hand cancer survivors may still entrust and may insist to include food supplements in order to enhance their overall health with an incipient feeling of cancer cure (Rock et al. 2004; Bours et al. 2015). The research in the area of food–drug interaction has demonstrated that the cancer patients who are experiencing the higher doses of supplement may adversely affect the efficacy of cancer chemotherapy drugs and may also create long-term health adversities (American Cancer Society Medical and Editorial Content Team 2015; Chen et al. 2019). Therefore, it is important to consider the quantification of required dose of dietary supplement to be consumed by the cancer patients, and investigate the contribution of the supplements to the total intake of nutrients in cancer survivors.

In the USA, previous cohorts study conducted with breast, prostate and colorectal cancer patients revealed higher frequency (50–85%) of DS usage (Rock 2007; Miller et al. 2008, 2009). A recent study conducted to understand the use of DSs among adult cancer survivors in the USA revealed interesting results describing the usage of DSs among cancer survivors with respect to their total nutrient intake as compared to the individuals who are without cancer. The study analysed the dose, prevalence and reason for consuming DSs among 2772 adult cancer patients and over 31,310 individuals without cancer, which participated in the National Health and Nutrition Examination Survey (NHANES) 2003–2016. It was observed that cancer survivors had a higher prevalence of any (70.4% vs. 51.2%) and multivitamin/mineral (48.9% vs. 36.6%) supplement consumption along with intake of 11 different vitamins and 8 minerals as compared to normal individuals (without cancer). The overall observation suggests that cancer patients had substantially higher nutrient intake from supplements, but comparatively much lesser nutrient intake from diet and foods. Interestingly, cancer patients had a higher percentage of individuals with inadequate intake for folate, niacin, vitamin B-6, calcium, phosphorus and copper owing to lower intakes of these nutrients from dietary sources as compared to normal individuals. Ironically, the cancer survivors also had a large number of individuals who had excess consumption of vitamin B-6, vitamin D, niacin, magnesium, calcium and zinc, which are achieved by higher consumption of DSs. Surprisingly, over half (46.1%) individuals used DSs on self-prescription and even without consulting physicians or healthcare providers (Du et al. 2020). Other study group analysed the NHANES trial data from 1999 to 2010, comprising of over 30,899 US adults aged 20 years, and reported that excess consumption of calcium supplements was significantly associated with increased risk of cancer death. However, the study finally concluded that the consumption of dietary supplements was not associated with mortality benefits among US adults (Chen et al. 2019).

Mitigating the "Profit Rich-Benefits Poor" Market of Dietary Supplements and Educating the Cancer Survivors

Besides the ground clinical realities about the benefits of DSs in cancer prevention, the market of DSs is growing in logarithmic fashion. US people spent over $36 billion in 2014 on dietary supplements. In fact, the consumer victimizing advertisements of DSs are easily accessed by cancer survivors through Internet, social media and television and inspire them to use it as an adjunct or traditional therapy. These self-prescriptions of DSs by cancer patients are mostly based on anecdotal evidences. More often owing to the drastic change in the psychology and mindset of cancer patients, they do not inform the inclusions of DSs to oncologist or their primary caretakers. In the USA, the regulatory affairs of DSs come under Dietary Supplement Health and Education Act of 1994 (DSHEA-1994), and after 9 years of DSHEA implementation, a panel of pharmacologists assessed the health information of all websites with reference to eight most widely used herbal supplements such as Ginkgo biloba, St John's wort, ginseng, garlic, echinacea, saw palmetto, valerian root and kava kava. The team observed that over 76% of retail sites were either selling DSs or had a linkage with vendors, and over 81% of retail sites had made one or multiple health claims. Among the total reviewed health claims, over 55% were to "treat", "cure-specific disease", "prevent" or "diagnose" mostly without the required

FDA disclaimer stating that the health claims had not been verified (Morris and Avorn 2003).

The Internet-based marketing does not seem to be restricted by federal laws. A case study of pomegranate juice/extract for prostate cancer prevention and treatment is an excellent example of how negative results of the clinical trials were marketed in DS market through Internet promotions. Prostate cancer patient searching on Google comes across the benefits of consuming pomegranate juice as a cancer-preventing DS. For instance, if you make a Google search for "pomegranate and cancer" one can identify over one million websites. Surprisingly, among the top four sites identified, three are reputed medical centres and one of them highlighted "The powerful pomegranate", "Scintillating cinnamon", "The mighty mushroom", "Brilliant blueberries" and "Glorious grapes" as "superfood" for cancer prevention by citing their own research without convincing clinical evidence (available online at http://nationalevents.cityofhope.org/site/PageNavigator/walk_super_foods). The description is more like victimizing customers. One more site exhibited a published review stating that consumption of pomegranate juice increased the mean PSA-DT (prostate-specific antigen doubling time) levels from 15 months at baseline to 54 months after treatment in men with up-surging PSA levels after surgery or radiation treatment (Pantuck et al. 2006). There were several lacunae in the review article, and one of the most important lacunae was that the authors did not mention that even though the PSA-DT values are closely correlated with progression-free and overall survival in prostate cancer, however, the change in PSA-DT cannot be considered as a primary endpoint owing to its uncertain clinical significance, and this is one of the key findings obtained after series of studies conducted by the Prostate Cancer Working Group (Scher et al. 2008). Besides these scientific shortcomings and having no demonstrable of anticancer activity, the Internet-driven promotion of these POM Wonderful pomegranate products resulted skyrocketing of sales, which were less than $12 million per year (9 years before) to over $150 millions in 2012 (Chaston 2015). In 2010, the US Federal Trade Commission (FTC) registered a complaint against POM Wonderful product stating that the antiprostate cancer claims of the manufacturer were fake, unsubstantiated and neither controlled nor blinded. Once the FTC administrative judge supported the findings along with support of US Court of Appeals to FTC's decision (POM Wonderful, LLC v Federal Trade Commission 2014), the POM Wonderful stopped advertising the prostate cancer benefits in its advertisements.

It is generally observed that the self-prescription intake of DSs by cancer survivors many times cross the recommended limits of its intake. The outcome of the NHANES 2003–2016 analysis revealed that approximately half of the cancer patients consumed DSs on their self-prescription, without consulting their healthcare personals. This situation also reflects the communication gap among the cancer patients and their respective healthcare agencies regarding DS use during clinical visits (Du et al. 2020). In a survey type study, it was observed that only 2% of the 1081 cancer patients consulted registered dietician towards the use of DS. As expected, the main reason for cancer survivors to use dietary supplements was to uplift and sustain overall health and quality of life (Pouchieu et al. 2015), and in this case, the use of DSs among such cancer patients was positively related to educational awareness, high income and an overall healthy lifestyle (Ferrucci et al. 2009).

The unrealized harm and risk of such high doses of DSs have been reported by series of RCTs and prospective cohort studies. For example, the "Alpha-Tocopherol, Beta-Carotene Cancer Prevention" (ATBC) trail and the "Beta-Carotene and Retinol Efficacy Trial" (CARET) found that supplemental use of β-carotene (20 or 30 mg/day) significantly escalated the risk of lung cancer among the smokers (Alpha-Tocopherol, Beta-Carotene Cancer Prevention Study Group 1994; Omenn et al. 1996). Almost similar trend was observed in "The Selenium and Vitamin E Cancer Prevention Trial" (SELECT), and the study observed that supplemental use of vitamin E (400 IU/day) significantly escalated the risk of prostate cancer among men (Klein et al. 2011). Yet, another observational study "Cancer Prevention Study" (CPS) II Nutrition Cohort reported that higher doses of supplemental calcium intake ($\geq$1000 mg/day) were strongly correlated with an elevated risk of all-cause mortality in men; however, the lower doses ($<$1000 mg/day) or calcium intake from dietary sources were not found to be related to mortality (Yang et al. 2016). Moreover, it was found that among the general US adult population who participated in the NHANES during 1999–2010, higher intake of calcium supplements ($\geq$1000 mg/day) was correlated with an increased cancer mortality (Chen et al. 2019).

The first important precaution that needs to be considered by cancer survivors is to assure the clinical reliability of the Internet-motivated claims of the DSs and not to become a victim of advertised false claims. The first-hand information in this regard can be retrieved from the "Clinicaltrials.gov" where over one thousand clinical trials of DSs have been registered. A Google advanced search on 10 November 2019 with a keyword "supplement" in Interventions field and "cancer" keyword in conditions field at clinicaltrials.gov displays more than 1100 registered clinical trials out of which 561 studies were completed. Replacing the keyword "supplement" by vitamins, the results show the registry of over 1200 studies out of which 561 are completed and accordingly several clinical trials are completed for tea (46), soya (42), selenium (32), lycopene (20), grapes (9), pomegranates (6) and resveratrol (5). The basic mandate of these registered trials is to provide evidence-based information to general public, oncologists and physicians, and moreover, the cancer survivors can find answers to their questions in relation to safety and efficacy of the DSs and FFs suggested for their possible implications in cancer prevention. The other reliable sources describing the detailed positive and negative impact of DSs (antioxidants, vitamin D, calcium, tea, etc.) are available online at the NCI web page (https://www.cancer.gov/about-cancer/causes-prevention/risk/diet). Two more sites of NCI, the first one available at http://www.cancer.gov/about-cancer/treatment/cam/hp provides the information about the complementary and unconventional medicine for health professionals and the second is available at http://www.cancer.gov/about-cancer/treatment/cam/patient, describe the complementary and alternative medicine for cancer patients. The information available at ACS website may help cancer patient to know more about DSs so that they can make a more informed decision about using them safely (https://www.cancer.org/treatment/treatments-and-side-effects/complementary-and-alternative-medicine/dietary-supplements.html).

Perhaps the global agencies dedicated for cancer welfare like WCRF-AICR have recently published dietary recommendations for 18 different cancers in their CUP report 2018, which also provided information on DSs (https://www.wcrf.org/dietandcancer/recommendations-about). The recently published ASC guidelines for diet and physical activity for cancer patients have once again made an appeal of judicious usage of dietary supplements.

The ACS panel has described that there are several products that are frequently marketed under brand name of "Dietary Supplements" but in real sense are not truly "dietary" because several ingredients of such supplements are other than food sources and are not found in foods consumed by humans. Nevertheless, even they cannot be described as "supplemental" in nature as they are not found to increase the intake of micronutrients, which have been scientifically proved to be indispensable for human health. Moreover, current legitimate regulations and laws do not have any vigilance over such products and do not promise any guarantee that the so-called dietary supplements truly contain the ingredients in the claimed quantities that are mentioned on their pamphlets or the products do not contain any substances that may directly or indirectly can be harmful to human health (Rock et al. 2020).

It is also worthy of visiting the websites of several national and international agencies of cancer welfare who have provided dietary guidelines along with recommendations for DSs (Table 14.1).

Conclusion

In an effort to get rescue from the drudgery of cancer, many cancer patients frequently consume variety of DSs. In fact, in the present state of the art, linking the pathophysiological relationship between various DSs and cancer prognosis or cure is a complicated issue, as outcomes of DSs are likely to be governed by several factors such as overall lifestyle quirks, immune status, type and stage of cancer, chemotreatment prescribed by oncologists, type, doses, and formulations of DSs, duration of DS consumption, etc. What is most important is cancer survivors should be encouraged and inspired to achieve micronutrients and phytonutrients by ingesting a diet including ample amounts of fruits, vegetables, legumes, nuts and seeds while avoiding the DSs intake. Apart from diet, equally important is keeping a healthy lifestyle, which is directly linked to lowering the probability of developing cancer and/or any other chronic diseases. Even as per the guidelines of series of national cancer organizations that support the usage of DSs, usually does not exceed 100% of the recommended daily allowance when dietary nutrient intake is severely inadequate or nutrition deficiencies are demonstrated (Luo and Asher 2017). Therefore, it is always imperative to consult a scientifically trained and accomplished person who can indulge proper council concerning intake of DSs.

Cancer survivors are advised to refer DS-related information from worldwide cancer-related organizations such as WCRF/AICR, ACS, NCI and NCCN, which mostly provides the authentic and more scientific guidelines towards consumption of

Table 14.1 Comparative profile of dietary supplement recommendations by national and international agencies for cancer patients

Name of the organization	Dietary supplements, micronutrients and vitamins	Online resources[a]
WCRF-AICR	• Do not use high-dose dietary supplements for cancer prevention • Avoid micronutrient supplement as it can be achieved through diet	• www.wcrf.org/sites/default/files/Recommendations • Continues Update Project Expert Report 2018. Available at dietandcancerreport.org • "HEAL Well: A Cancer Nutrition Guide" by AICR. Available at www.aicr.org/assets/docs/pdf/education/heal-well-guide.pdf
NCI	• Vitamin C, probiotics, melatonin, oral glutamine	• www.cancer.gov/publications/patient-education/eatinghints; • www.cancer.gov/about-cancer/treatment/side-effects/appetite-loss/nutrition-pdq • www.cancer.gov/about-cancer/causes-prevention/risk/diet
ACS	• Eat dietary antioxidants and not as supplements • Eat recommended levels of calcium and vitamin D • Intake of antioxidant supplements or vitamin-enhanced foods is not recommended while chemo- or radiation therapy	www.cancer.org/treatment/survivorship-during-and-after-treatment
WHO: ECAC	• Dietary supplements are not recommended for cancer prevention • It is best to have a diet of ordinary foods that will provide all the nutrients	http://cancer-code-europe.iarc.fr/index.php/en/ecac-12-ways/diet-recommendation
NCCN	• Food—not supplements—is the best source of vitamins and minerals • Eat foods high in vitamin D (salmon, sardine, fortified orange juice, milk, etc.)	www.nccn.org/patients/resources/life_after_cancer/nutrition.aspx
ESPEN	• Oral nutritional supplements like multivitamins and fish oil supplements (advanced cancer) are recommended to avoid nutritional deterioration	www.espen.org/guidelines-home/espen-guidelines
CRUK	• Supplements do not substitute for a healthy diet, so eat a healthy and balanced diet • Everyone should consider taking vitamin D	www.cruk.org/diet
CSC	• Do not use supplements for cancer prevention. A healthy diet is the best way to get nutrition	www.cancersupportcommunity.org/diet-nutrition

(continued)

Table 14.1 (continued)

Name of the organization	Dietary supplements, micronutrients and vitamins	Online resources[a]
CCS	• Commercial nutrition supplements of vitamins, minerals and others are recommended, which is right for patients	www.cancer.ca/en/cancer-information/cancer-journey/living-with-cancer/nutrition-for-people-with-cancer
WHO	The WHO has established Vitamin and Mineral Nutrition Information System (VMNIS) to strengthen the global surveillance of vitamin and mineral deficiencies	• www.who.int/en/news-room/fact-sheets/detail/healthy-diet • Healthy diet-2019 (https://apps.who.int/iris/handle/10665/325828) • www.who.int/features/qa/cancer-red-meat/en/

WHO World Health Organization, *WCRF-AICR* World Cancer Research Fund–American Institute for Cancer Research, *NCI* National Cancer Institute, *ASN* American Society for Nutrition, *ACS* American Cancer Society, *CSC* Cancer Support Community, *CCS* Canadian Cancer Society, *WHO: ECAC* European Code Against Cancer, *NCCN* National Comprehensive Cancer Network, *ESPEN* The European Society for Clinical Nutrition and Metabolism, *CRUK* Cancer Research UK
[a]Accessed in September 2020

DSs. It is always advisable, not to be get victimized towards the "Profit Rich-Benefits Poor" advertisements of profit earning agencies.

Because of the adverse effects of the chemotherapy and restricted diet of many cancer patients, the inclusion of multivitamins or multimineral supplements in doses close to the recommended dietary limits under the supervision of medical professionals may help cancer patients; however, the judicious decision of oncologists in concert with stage of cancer progression, age of the cancer patient, the vitamin or mineral deficiency profile of the cancer survivor and the dose of DSs are some of the factors associated with efficacy of DSs in cancer management (Ravasco 2019).

Looking forward towards the judicious use of DSs, perhaps conducting double-blinded randomized controlled trials (RCTs) seems to be a gold standard for integrating the rational usage of DSs in clinical practice. RCTs are important because many times the findings from the observational trials do not mount to clinical reality. Therefore, design and conduct of RCTs should be the state-of-the-art methodology for getting a clearer picture of the pathophysiological role of DSs in prevention and treatment of cancer. Nevertheless, the trial must enrol a diverse population of individuals who are actually highly susceptible to develop a nutritional deficiency or individuals that have already developed deficiency of nutrients, as several trials have enrolled primarily women and men healthy volunteers, which are highly educated participants.

References

Alpha-Tocopherol, Beta-Carotene Cancer Prevention Study Group. The effect of vitamin E and beta carotene on the incidence of lung cancer and other cancers in male smokers. N Engl J Med. 1994;330(15):1029–35.

American Cancer Society Medical and Editorial Content Team. Risks and side effects of dietary supplements: American Cancer Society; 2015 [updated 31 Mar 2015]. https://www.cancer.org/treatment/treatments-and-side-effects/complementary-and-alternative-medicine/dietary-supplements/risks-and-side effects.html. Accessed 26 Sept 2020.

Barnes K, Ball L, Desbrow B, et al. Consumption and reasons for use of dietary supplements in an Australian university population. Nutrition. 2016;32:524–30.

Bours MJ, Beijer S, Winkels RM, et al. Dietary changes and dietary supplement use, and underlying motives for these habits reported by colorectal cancer survivors of the Patient Reported Outcomes Following Initial Treatment and Long-Term Evaluation of Survivorship (PROFILES) registry. Br J Nutr. 2015;114(2):286–96.

Chaston I. Entrepreneurial marketing: sustaining growth in all organisations. London: Palgrave Macmillan; 2015.

Chen F, Du M, Blumberg JB, et al. Association among dietary supplement use, nutrient intake, and mortality among U.S. adults: a cohort study. Ann Intern Med. 2019;170(9):604–13.

Cowan A, Jun S, Gahche J, et al. Dietary supplement use differs by socioeconomic and health-related characteristics among U.S. adults, NHANES 2011–2014. Nutrients. 2018;10:1114.

Czerwiński A, Liebers D. Poland and supplements: opportunities for European leadership. Warsaw: Polish Economic Institute; 2019.

Dickinson A, MacKay D. Health habits and other characteristics of dietary supplement users: a review. Nutr J. 2014;13:14.

Du M, Luo H, Blumberg JB, et al. Dietary supplement use among adult cancer survivors in the United States. J Nutr. 2020;150(6):1499–508.

Ferrucci LM, McCorkle R, Smith T, Stein KD, Cartmel B. Factors related to the use of dietary supplements by cancer survivors. J Altern Complement Med. 2009;15(6):673–80.

Hys K. Identification of the reasons why individual consumers purchase dietary supplements. In: Sroka W, editor. Perspectives on consumer behaviour. Theoretical aspects and practical applications. Cham: Springer; 2020. p. 193–209.

Kim Y, Je Y. Vitamin D intake, blood 25(OH)D levels, and breast cancer risk or mortality: a meta-analysis. Br J Cancer. 2014;110(11):2772–84.

Klein EA, Thompson IM Jr, Tangen CM, et al. Vitamin E and the risk of prostate cancer: the Selenium and Vitamin E Cancer Prevention Trial (SELECT). JAMA. 2011;306(14):1549–56.

Luo Q, Asher GN. Complementary and alternative medicine use at a comprehensive cancer center. Integr Cancer Ther. 2017;16(1):104–9.

Miller P, Demark-Wahnefried W, Snyder DC, et al. Dietary supplement use among elderly, long-term cancer survivors. J Cancer Surviv. 2008;2(3):138–48.

Miller PE, Vasey JJ, Short PF, Hartman TJ. Dietary supplement use in adult cancer survivors. Oncol Nurs Forum. 2009;36(1):61–8.

Morris CA, Avorn J. Internet marketing of herbal products. JAMA. 2003;290(11):1505–9.

Omenn GS, Goodman GE, Thornquist MD, et al. Effects of a combination of beta carotene and vitamin A on lung cancer and cardiovascular disease. N Engl J Med. 1996;334(18):1150–5.

Pantuck AJ, Leppert JT, Zomorodian N, et al. Phase II study of pomegranate juice for men with rising prostate-specific antigen following surgery or radiation for prostate cancer. Clin Cancer Res. 2006;12(13):4018–26.

Patterson RE, Neuhouser ML, Hedderson MM, Schwartz SM, Standish LJ, Bowen DJ. Changes in diet, physical activity, and supplement use among adults diagnosed with cancer. J Am Diet Assoc. 2003;103(3):323–8.

POM Wonderful, LLC v Federal Trade Commission. US Court of Appeals. Argued 2 May 2014. Available online.

Pouchieu C, Fassier P, Druesne-Pecollo N, et al. Dietary supplement use among cancer survivors of the NutriNet-Sante cohort study. Br J Nutr. 2015;113(8):1319–29.

Ravasco P. Nutrition in cancer patients. J Clin Med. 2019;8(8). pii: E1211.

Rock CL. Multivitamin-multimineral supplements: who uses them? Am J Clin Nutr. 2007;85 (1):277S–9S.

Rock CL, Newman VA, Neuhouser ML, Major J, Barnett MJ. Antioxidant supplement use in cancer survivors and the general population. J Nutr. 2004;134(11):3194s–5s.

Rock CL, Doyle C, Demark-Wahnefried W, Meyerhardt J, et al. Nutrition and physical activity guidelines for cancer survivors. CA Cancer J Clin. 2012;62(4):243–74.

Rock CL, Thomson C, Gansler T, et al. American Cancer Society guideline for diet and physical activity for cancer prevention. CA Cancer J Clin. 2020;70(4):245–71.

Satia JA, Walsh JF, Pruthi RS. Health behaviour changes in white and African American prostate cancer survivors. Cancer Nurs. 2009;32(2):107–17.

Scher HI, Halabi S, Tannock I, et al. Prostate Cancer Clinical Trials Working Group. Design and end points of clinical trials for patients with progressive prostate cancer and castrate levels of testosterone: recommendations of the Prostate Cancer Clinical Trials Working Group. J Clin Oncol. 2008;26(7):1148–59.

Sekhri K, Kaur K. Public knowledge, use and attitude toward multivitamin supplementation: a cross-sectional study among general public. Int J Appl Basic Med Res. 2014;4:77.

Trujillo EB, Claghorn K, Dixon SW, Hill EB, Braun A, Lipinski E, Platek ME, Vergo MT, Spees C. Inadequate nutrition coverage in outpatient cancer centers: results of a national survey. J Oncol. 2019;2019:1–8.

Velicer CM, Ulrich CM. Vitamin and mineral supplement use among US adults after cancer diagnosis: a systematic review. J Clin Oncol. 2008;26(4):665–73.

Yang B, Campbell PT, Gapstur SM, et al. Calcium intake and mortality from all causes, cancer, and cardiovascular disease: the Cancer Prevention Study II Nutrition Cohort. Am J Clin Nutr. 2016;103(3):886–94.

Yao S, Kwan ML, Ergas IJ, et al. Association of serum level of vitamin D at diagnosis with breast cancer survival: a case-cohort analysis in the pathways study. JAMA Oncol. 2017;3(3):351–7.

Zhang FF, Liu S, John EM, Must A, Demark-Wahnefried W. Diet quality of cancer survivors and noncancer individuals: results from a national survey. Cancer. 2015;121(23):4212–21.

Zhang FF, Barr SI, McNulty H, et al. Health effects of vitamin and mineral supplements. BMJ. 2020;369:m2511.

Chapter 15
Cancer and Fasting: Can Fasting/Calorie Restriction (CR) or Fasting-Mimicking Diet (FMD) Help Fight Cancer? Molecular Basis of Fasting Response

Abstract The ecosystem of cancer management is exploring the multiple approaches for the effective management of cancer. In the quest of integrated cancer management, the beneficial therapeutic effects of fasting/calorie restriction (CR) or fasting-mimicking diet (FMD) have inspired researchers and clinicians owing to their benefits to cancer patients towards overall survival, disease prevention, improvements in treatments and amelioration of side effects and off-target toxicities. There are growing preclinical clinical evidences, which describe the anticancer effects of fasting/CR or FMD by reducing cancer progression, inducing the mortality of cancer cells and improving the effectiveness and tolerability of chemo- and radiotherapies. Series of clinical trials are being launched and several completed (ClinicalTrials.gov) for addressing the interventions of fasting/CR or FMD in caner management; however, the overall outcome of studies seems to be inconsistent and perhaps focused randomized clinical trials are still required to understand their multidimensional molecular effects under different settings of cancer therapy. As a part of understanding the molecular basis of fasting/CR or FMD, the experimental data indicate that fasting/CR or FMDs significantly deplete the levels of tumour growth essential nutrients and factors, including glucose, IGF1 and insulin. Apart from significant decrease in the ratio of ATP/ADP and NADH/NAD, there is upregulation of variety of proteins, which are involved in apoptosis, autophagy, detoxification, DNA repair, genomic stability and immunosurveillance, while several proteins and pathways implicated in cell proliferation and oxidative damage are downregulated. Interestingly, fasting/CR or FMD may demonstrate altogether different and even opposite effects in different cancer cell types or even within the same cancer cell type.

Keywords Fasting · Calorie restriction · Fasting-mimicking diet · Cancer prevention · Chemotherapy · Molecular mechanism

R. N. Gacche, *Dietary Research and Cancer*,
https://doi.org/10.1007/978-981-16-6050-4_15

Abbreviations

AKT	Protein kinase B
AMPK	AMP-activated protein kinase
Bcl-xL	B-cell lymphoma extra large
cAMP	Cyclic adenosine monophosphate
CD	Cluster of differentiation
CR	Calorie restriction
FOXO	Forkhead box protein O
GLUT	Glucose transporter
HDAC	Histone deacetylase
HIF-1α	Hypoxia-inducible factor 1-alpha
IGF	Insulin-like growth factor
IGF-R	Insulin-like growth factor 1 receptor
MAPK	Mitogen-activated protein kinase
mTOR	Mammalian target of rapamycin
mTORC1	Mammalian target of rapamycin complex 1
NF-κB	Nuclear factor-kappa B
Nrf2	Nuclear factor erythroid-2-related factor 2
p53	Tumour suppressor protein that protects from DNA damage
P70S6K	Ribosomal protein S6 kinase
PI3K	Phosphoinositide 3-kinase
PTEN	Phosphatase and tensin homolog
ROS	Reactive oxygen species
SIRT-1	Sirtuin-1
SOD	Superoxide dismutase

Introduction

There is growing curiosity among the cancer survivors towards an unresolved question of whether fasting/CR/FMD has any role to play in the progression and treatment of cancer. Few notable questions being asked by the cancer patients pertain to efficacy of fasting/CR/FMD in arresting the progression of cancer, killing of cancer cells, boosting of the immune system and whether it has role in improving the effectiveness of currently prescribed chemotherapy and radiation therapy? In fact, it is well-established fact that dietary interventions and lifestyle-related factors are associated with risk of developing cancer in general and certain cancers being more influenced by dietary factors (Kushi et al. 2012; Calle et al. 2003; Emmons and Colditz 2017). Nevertheless, fasting has remained a integral part of various spiritual traditions across the world, most notably in religions like Hinduism, Christianity, Judaism, Islam and Buddhism. There are also glimpses in the history that many great spiritual leaders like Jesus, Buddha and Mohammed fasted for achieving spiritual and mental understanding. Human race is also sensitized for starvations owing to famines in many parts of the world. In general, it appears that most humans can sustain their life on water alone for over months. On evolutionary grounds, almost all organisms have survived in scanty food environments, during which starvation was an unavoidable and common condition, and perhaps evolution has favoured organisms that can sustain in starvation niche.

Before venturing into a discourse of "CR/fasting/FMD-cancer dynamics", it will be imperative to clearly define the terms related to fasting/CR/FMD. CR is related to limiting the consumption of food (20–40% cut down in calorie intake), avoiding the malnutrition, and fasting on the contrary is a voluntarily decision of not consuming any kind of food for varying time periods. Intermittent CR pertains to 50–70% cut down in calorie intake for a brief time period; however, intermittent fasting means absolute avoidance of calorie intake daily for 16–18 h or alternate-day fasting. Fasting should not be confused with CR, which is an entirely different pattern of eating, which includes a consistent reduction in average daily caloric intake below to what is typical or habitual, without deprivation of essential nutrients. FMD is a periodic fasting diet regime that initially was developed to test the efficacy of CR and fasting in animal models, and consequently, the same was extended for humans. Actually in 2015, Brandhorst and his team tested this hypothesis in vitro on mice, and they designed and refined a feeding protocol that could mimic the effects of fasting, and they termed it as FMD. This diet consists of a sum total energy intake restricted between 10 and 50% of the diet ad libitum, and a low-protein, a low-carbohydrate and moderately high-fat composition that would last for 4 days. The results and outcomes of this study were very interesting, and they displayed that the mice which were fed according to this diet for 4 days consecutively demonstrated similar outcome, especially in terms of weight loss and improvement in circulating metabolic risk factor as compared to short-term starvation for 3 days (Brandhorst et al. 2015).

Fasting has its reputation as one of the older most traditional medicinal therapies across the world. Since almost over period of 100 years, the scientific community is trying to decipher the role of calorie restriction in connection to its possibility of prolonging the life. Majority of the research has been performed in preclinical settings, and a modest volume of literature has accumulated in the recent past that indicates the protective effect of CR and fasting against secondary ageing. Moreover, CR downregulated the levels of a number of anabolic hormones, inflammatory cytokines and growth factors, reduces cell proliferation and oxidative stress and, interestingly upregulates the autophagy (self-eating of cell) and several DNA repair processes (Longo and Fontana 2010). Although sizable literature has accumulated in the recent past linking the role of fasting or CR in cancer progression, however, there is substantial scope and opportunity for the upcoming or ongoing clinical trials to unravel and decipher whether the fasting or CR really works against cancer progression? Whether it ameliorates the off-target toxicity and boosts the efficacy of chemotherapeutic regimens in daily clinical practice? The unravelling of the molecular underpinning of above questions might pave way for practical utility of fasting or CR in the management of cancer, which otherwise is limited due to lack of clinical evidence.

Before we venture into the molecular aspects of fasting on cancer progression, let us refresh the classical understanding of metabolic changes that undergo in mammals after fasting or CR. Mammals undergo series of metabolic perturbations when deprived from food. Once we eat the food, after the postabsorptive phase, which usually lasts for 10 or more hours, primarily the utilization of glycogen takes place as the main source of stored energy. After the exhaust of the hepatic glycogen reserves, amino acid pool serves as gluconeogenic substrates, while most of the leftover glucose in the body is utilized by the brain. After about 72 h of consistent fasting, the blood glucose levels may reach up to 50–60 mg/dL in a normal person; however, by the oral intake of 100 g glucose, it can be restored to normal levels within a span of 30 min. Further after a few days of fasting the fatty acids and glycerol and are released from the adipose tissue, and consequently, the acetoacetate, b-hydroxybutyrate and acetone (ketone bodies) become the primary source of carbon for tissue survival. In the final phase of prolonged fasting, after the fat storage has been exhausted, the muscles undergo degradation, which drives the process of gluconeogenesis (generation of sugar from noncarbohydrate precursors). For mammals, especially in humans, it is estimated that during fasting a healthy individual having weight of ~70 kg can achieve basal caloric requirements from stored fats up to 2–3 months, which eventually leads to initially rapid weight loss that eventually tapers off. In general, in the first week of fasting, on an average 0.9 kg/day is depleted, followed by a depletion of 0.3 kg/day by the third week of fasting, resulting in an approximately 20% body weight loss after 30–35 days of fasting (Brandhorst and Longo 2016).

Looking forward towards the multidimensional effects of fasting/CR and FMD, the most upcoming discourse will be to test their intervention in regulating the dynamics of cancer microbiome. In the midst of evolving human microbiome projects across the world, it has been described that the human body has more microorganism than human cells (Sender et al. 2016).

The past decades of research inspired the research community to infer the disease biology with human microbiome owing to groundbreaking advances in the field of microbiome research. Sizable literature is available, which describes the inscrutable links of the dynamics of gut microbiota with pathophysiological complexities of tumour biology. It has been demonstrated that the microbiome plays a role in tumour initiation and progression and influence directly on the tumour cells and indirectly via the manipulation and regulation of the immune system. The recent literature also showed that dynamics of microbiome also determines the response to cancer therapies, and predicts progression of disease and patient survival. Moreover, modulation of the microbiome can be utilized to explore the efficacy of immunotherapies and for amelioration of drug toxicity (Jain et al. 2021). The logic of linking fasting/CR and FMD with microbiome lies in the deprivation of nutrients to the microbial community and understanding their effects on cancer progression and treatments. It is interesting to note that the microbiome is inevitably amalgamated into the cancer–immune crosstalk and coordination. The gut microbes are known to interact with the host and influence the tumour microenvironment in a diversified and context-specific manner, conferring individual and tumour specificities to the immunomodulatory phenotype observed. Unravelling the unique molecular mechanisms through which these alterations are manifested may help in designing personalized and targeted approaches for cancer diagnosis, treatment and prevention (Fessler et al. 2019). The era of immuno-onco-microbiotics is extremely exciting and the excitement can be increased many folds by studying the role of fasting, CR and FMD in regulation of cancer-associated microbiome.

Molecular Basis of Fasting Response

The evolving preclinical research to a greater extent and the clinical research to a lesser extent outline the susceptibility of cancer cells to nutrient deprivation and their dependence on specific metabolites as emerging hallmarks of cancer. Implications of fasting or fasting-mimicking diets (FMD) perturb variety of changes in growth factors and levels of metabolites, thereby mustering environments that can hinder the capability of cancer cells to adapt and survive and perhaps may supplement in improving the effects of cancer therapeutic modalities. Moreover, the research findings also suggest that fasting or CR may increase resistance to chemotherapy in normal cells but not in cancer cells and thereby promote regeneration/remodelling in normal tissues, which may help in preventing detrimental and deleterious life-threatening side effects of cancer treatments. In the current state-of-the-art literature, the dietary interventions based on fasting and CR have been extensively inspected in the mainstream of oncology, both under preclinical and clinical settings. The research also investigated the effect of water fasting (devoid of any food and drinks except water) and FMDs (Nencioni et al. 2018). Preliminary clinical observations also pointed out that a fasting of minimum of 48 h might result in meaningful clinical outcomes in oncology like inhibiting the chemotherapy-related DNA damage in normal tissues and may help in maintaining the good quality of life of cancer patient during chemotherapy (de Groot et al. 2015).

In fact, acclimatization to starvation condition compels the organism to channel energy into multiple protective processes that may scale down the physiological metabolism. It is also believed that these systems may prolong life and thereby decrease the incidence risk of cancer. It has been described that intervention of CR without leading to malnutrition is the most impressive and authentic reproducible physiological intervention for increasing the lifespan and providing protection against cancer in mammalian animals (Longo and Fontana 2010). More precisely, the fasting or CR perturbs the circulating levels of growth hormone, glucose, insulin, insulin growth factor 1 (IGF1), glucagon, (GH), adrenaline and glucocorticoids. In the initial phase, i.e. postabsorptive phase of fasting (6–24 h), the insulin levels start to decline; on the other hand, the level of glucagon rises and accelerates the breakdown of liver glycogen and releases free glucose for energy production. Parallelly, GH and lowered levels of insulin further accelerate the breakdown of triglycerides leading to generation of glycerol and free fatty acids. During CR or fasting, most tissues consume fatty acids as a source of energy. On the contrary, the brain tissues rely on glucose and ketone bodies generated by hepatocytes. In short, the brain utilizes ketone bodies, glycerol and amino acids derived from fat (Nencioni et al. 2018). Finally, CR or fasting also regulates the levels of hormones related to digestion and metabolism, where it reduces level of leptin (an hunger-inhibiting hormone secreted by adipocytes), while it elevates levels of adiponectin (which is responsible for breakdown of fatty acid). Thus, in conclusion, the hallmarks of the mammalian systemic response to fasting are lowered levels of insulin and glucose, and elevated levels of glucagon and ketone bodies, coupled with lowered levels of IGF1 and leptin and elevated levels of adiponectin (Brennan and Mantzoros 2006; Kubota et al. 2007). Per se, ketone bodies are known to inhibit histone deacetylases (HDACs), and therefore, the fasting-stimulated overgenerated ketone bodies can attenuate the tumour growth and might also simulate differentiation through various epigenetic mechanisms (Newman and Verdin 2014). However, in certain melanomas with mutated B-RAF, the ketone body acetoacetate has been observed to expedite, instead of reducing, the tumour growth (Xia et al. 2017). The plausible mechanisms of fasting or FMD-induced mortality of cancer cells in solid tumours are shown in Fig. 15.1.

While depicting fasting or CR effects at cellular levels, in model organisms, it seems that the response of normal cells towards fasting is conserved evolutionarily and involved in conferring cell protection and thereby plays a role in increasing health span and lifespan (Brandhorst et al. 2015). Of the studied pathways, perhaps, IGF1 signalling cascade seems to be a key signalling pathway mediating the effects of fasting at the cellular levels. Under normal nutritional status, the consumed protein diet contributes towards increasing the levels of amino acids, which has been linked with upregulation of IGF1 levels, and also enhances the activity of AKT and mTOR proteins, which ultimately leads to accelerating the synthesis of proteins. On the contrary, during fasting state, the levels of IGF1 and related downstream signalling pathways are downregulated. This ultimately leads to reducing the AKT-mediated inhibition of mammalian FOXO transcription factors and thereby permitting these transcription factors to transactivate genes, leading to the upregulation of and activation of series of enzymes, i.e. superoxide dismutase (SOD), haem oxygenase 1 (HO1) and catalase, which have in-built antioxidant and protective effects (Cheng et al. 2009; Converso et al. 2006). Moreover, elevated levels of glucose activate protein kinase A signalling cascade, which adversely alters the regulation of the master energy sensor AMPK: AMP-activated protein kinase (Hurley et al. 2006) and consequently downregulates the stress resistance transcription factor such as EGR1: early growth response protein 1 (Berasi et al. 2006). It has been also reported that fasting-related glucose restriction inhibits the PKA activity, upregulates AMPK functions and activates EGR1, which ultimately confers the cell-protective effects, including the protection to myocardium (Brandhorst et al. 2015; Cheng et al. 2014; Di Biase et al. 2017). Studies have also demonstrated that fasting or fasting-mimicking diet (FMD) downregulates the intracellular signalling pathways such as the signalling axis IGF1R–AKT–mTOR–S6K and also cAMP–PKA

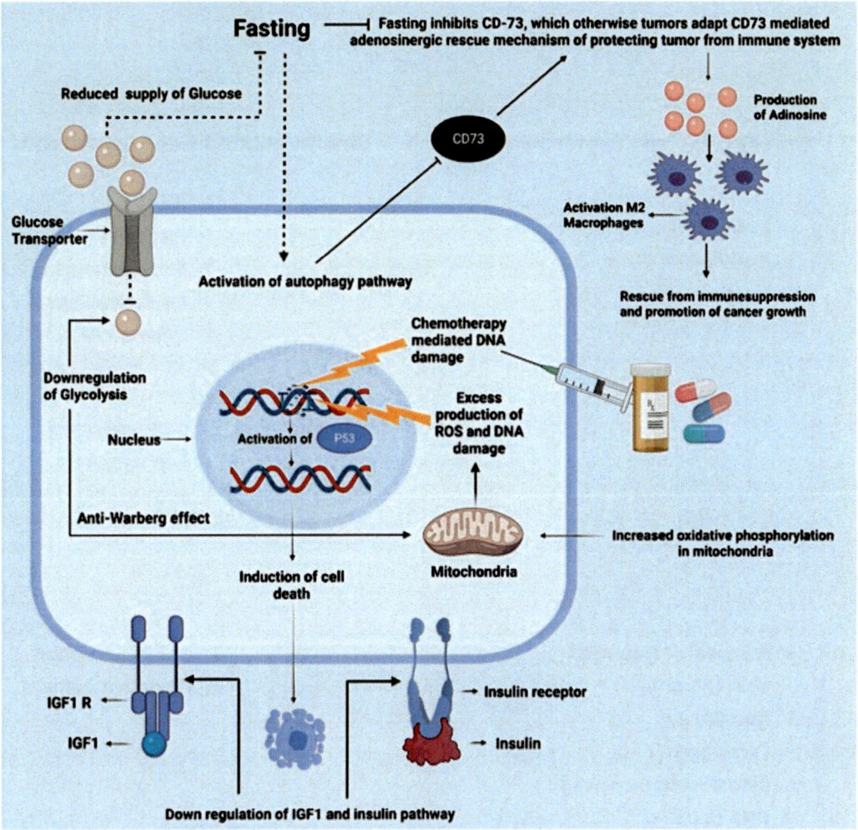

Fig. 15.1 Possible mechanisms of fasting, CR or FMD-induced mortality of cancer cells in solid tumours. The experimental data imply that fasting or fasting-mimicking diets (FMDs) significantly deplete the levels of tumour growth essential nutrients and factors such as glucose, insulin and IGF1. Fasting may also create an anti-Warburg effect by minimizing glucose uptake by downregulating the glucose transporters (GLUTs) and thereby aerobic glycolysis and thus compelling the tumours to upregulate the oxidative phosphorylation, which in turn leads to increase in the production of reactive oxygen species (ROS) in cancer cells. The generated oxidative stress damages DNA, activates p53 and compels the cells to undergo apoptotic cell death. The similar kind of oxidative stress is possibly generated by chemotherapy drugs and contributes towards cancer cell death. Fasting also activates autophagy and reduces the levels of CD73 in certain cancer cells, thus hampering the production of adenosine in the extracellular environment and thereby averting the macrophages transformation towards an immunosuppressive M2 phenotype. Interestingly, fasting or FMD may demonstrate altogether different and even opposite effects in different cancer cell types or even within the same cancer cell type

signalling cascade. The fasting or FMD is also found to upregulate the autophagy, equips the normal cells to withstand against stress and thereby promotes anticancer immunity (Cheng et al. 2014; Pietrocola et al. 2016; Di Biase et al. 2016). The effect of low-calorie intake/fasting on regulation metabolic, molecular signalling and

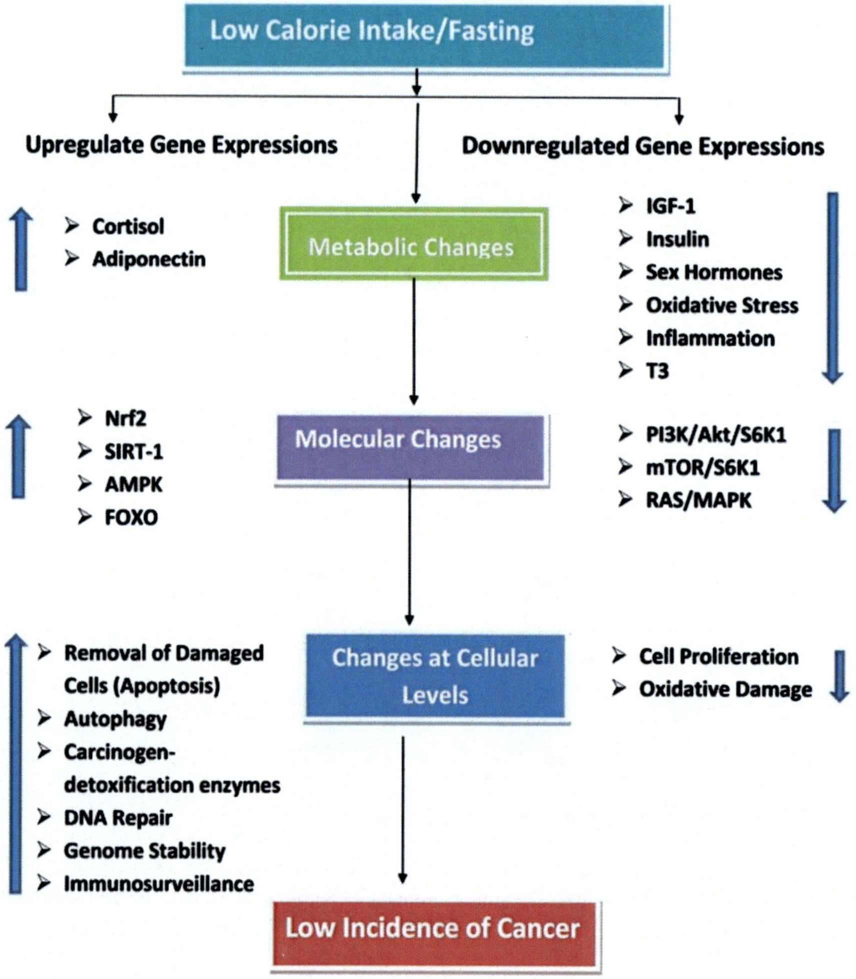

Fig. 15.2 Effect of fasting/low-calorie intake on regulation of metabolic, molecular and cellular processes leading to low incidence of cancer

cellular processes, which leads to low incidence of cancer, is summarized in Fig. 15.2.

Apart from significant decrease in the ratio of ATP/ADP and NADH/NAD, there is upregulation of variety of proteins, which are involved in apoptosis, autophagy, detoxification, DNA repair, genomic stability and immunosurveillance; while several proteins and pathways implicated in cell proliferation and oxidative damage are downregulated. Interestingly, fasting/CR or FMD may demonstrate altogether diverse and even conflicting effects in different cancer types or even within the same type of cancer.

The health benefits of fasting/CR and FMD can occur via circuitous impact on metabolism and body mass. For example, in the USA, the reports state that obesity and overweight are associated with over 14 and 20% of cancer-related deaths in males and females, respectively (Calle et al. 2003). Fasting/CR and FMD may also demonstrate effects on cancer cell growth factor, signalling pathways, hormone, inflammation, energy homeostasis and vascular changes, which may help in amelioration of the tumour microenvironment, which per se is a key factor in cancer progression and drug resistance. Nevertheless, fasting/CR or FMD may help also improve the tolerability and efficiency of cancer drugs (Nencioni et al. 2018). Perhaps these effects may get manifested through a decline in mitogenic stimuli in normal cells and hindering effect on the cell cycle in cancer cells (Lee et al. 2010; Raffaghello et al. 2008).

Conclusion

The evolving literature accumulated in the recent past clearly attributed the significance of fasting, CR and FMD or even biocompatible compounds mimicking the biological effects of fasting and CR (known as caloric restriction mimetics) with reducing the risk of cancer and other age-related degenerative diseases such as cardiovascular diseases, and cognitive disorders, and a role in extending the overall survival of cancer patients. There are growing preclinical clinical evidences, which describe the anticancer effects of fasting/CR or FMD by reducing cancer progression, inducing the mortality of cancer cells and improving the effectiveness and tolerability of chemo- and radiotherapies. Series of clinical trials are being launched and several completed (ClinicalTrials.gov) for addressing the interventions of fasting/CR or FMD in caner management; however, the overall outcome of studies seems to be inconsistent and perhaps focused randomized clinical trials are still required to understand their multidimensional molecular effects under different settings of cancer therapy. As a part of understanding the molecular basis of fasting/CR or FMD, the experimental data suggest that fasting/CR or FMDs significantly deplete the levels of tumour growth essential nutrients and factors, including glucose, IGF1 and insulin. Apart from significant decrease in the ratio of ATP/ADP and NADH/NAD, there is upregulation of variety of proteins, which are involved in apoptosis, autophagy, detoxification, DNA repair, genomic stability and immunosurveillance, while several proteins and pathways implicated in cell proliferation and oxidative damage are downregulated. Interestingly, fasting/CR or FMD may demonstrate altogether different and even opposite effects in different cancer cell types or even within the same cancer cell type.

Looking forward, the most upcoming discourse may be to investigate the dynamic interplay between fasting/CD/FMD and cancer microbiome. There is ample research opportunity to understand the diet-specific regulation of microbiome providing the benefits and risk to cancer patients. Moreover, the research opportunities in understanding the role of fasting/CR and FMD in improving cancer immunosurveillance and minimizing the migration of immunosuppressive cells (especially regulatory T cells) towards the tumour microenvironment may be useful in integrating these nonpharmacological regimens with cancer immune therapy. The current literature also sensitizes with the association of fasting/CR and FMD with health-supportive cancer microbiome and its dynamics. Nevertheless, fasting/CR and FMD may directly or indirectly affect the functional profile of carcinogenic and metastatic abilities of cancer stem cells, which are generally considered as a major culprit in recurrence, relapse and drug resistance. Finally, considering the scenario of promising beneficial effects of fasting CR and FMD, it will be indispensable to design and conduct controlled randomized large clinical trials for understanding the risk and benefits of fasting CR and FMD. Perhaps such endeavours may inspire the inclusion and integration of these nonpharmacological modalities in clinical practice for effective management of cancer.

References

Berasi SP, et al. Inhibition of gluconeogenesis through transcriptional activation of EGR1 and DUSP4 by AMP-activated kinase. J Biol Chem. 2006;281:27167–77.

Brandhorst S, Longo VD. Fasting and caloric restriction in cancer prevention and treatment. Recent Results Cancer Res. 2016;207:241–66.

Brandhorst S, et al. A periodic diet that mimics fasting promotes multi-system regeneration, enhanced cognitive performance, and health span. Cell Metab. 2015;22:86–99.

Brennan AM, Mantzoros CS. Drug insight: the role of leptin in human physiology and pathophysiology-emerging clinical applications. Nat Clin Pract Endocrinol Metab. 2006;2: 318–27.

Calle EE, Rodriguez C, Walker-Thurmond K, et al. Overweight, obesity, and mortality from cancer in a prospectively studied cohort of U. S. adults. N Engl J Med. 2003;348:1625–38.

Cheng Z, et al. Foxo1 integrates insulin signaling with mitochondrial function in the liver. Nat Med. 2009;15:1307–11.

Cheng CW, et al. Prolonged fasting reduces IGF-1/PKA to promote hematopoietic-stem-cell-based regeneration and reverse immunosuppression. Cell Stem Cell. 2014;14:810–23.

Converso DP, et al. HO-1 is located in liver mitochondria and modulates mitochondrial heme content and metabolism. FASEB J. 2006;20:1236–8.

de Groot S, et al. The effects of short-term fasting on tolerance to (neo) adjuvant chemotherapy in HER2-negative breast cancer patients: a randomized pilot study. BMC Cancer. 2015;15:652.

Di Biase S, et al. Fasting-mimicking diet reduces HO-1 to promote T cell-mediated tumor cytotoxicity. Cancer Cell. 2016;30:136–46.

Di Biase S, et al. Fasting regulates EGR1 and protects from glucose- and dexamethasone-dependent sensitization to chemotherapy. PLoS Biol. 2017;15:e2001951.

Emmons KM, Colditz GA. Realizing the potential of cancer prevention—the role of implementation science. N Engl J Med. 2017;376:986–90.

Fessler J, Matson V, Gajewski TF. Review exploring the emerging role of the microbiome in cancer immunotherapy. J Immunother Cancer. 2019;7(1):108.

Hurley RL, et al. Regulation of AMP-activated protein kinase by multisite phosphorylation in response to agents that elevate cellular cAMP. J Biol Chem. 2006;281:36662–72.

Jain T, Sharma P, Are AC, et al. New insights into the cancer-microbiome-immune axis: decrypting a decade of discoveries. Front Immunol. 2021;12:622064.

Kubota N, et al. Adiponectin stimulates AMP-activated protein kinase in the hypothalamus and increases food intake. Cell Metab. 2007;6:55–68.

Kushi LH, Doyle C, McCullough M, et al. American Cancer Society guidelines on nutrition and physical activity for cancer prevention: reducing the risk of cancer with healthy food choices and physical activity. CA Cancer J Clin. 2012;62(1):30–67.

Lee C, Safdie FM, Raffaghello L, et al. Reduced levels of IGF-I mediate differential protection of normal and cancer cells in response to fasting and improve chemotherapeutic index. Cancer Res. 2010;70(4):1564–72.

Longo VD, Fontana L. Calorie restriction and cancer prevention: metabolic and molecular mechanisms. Trends Pharmacol Sci. 2010;31:89–98.

Nencioni A, Caffa I, Cortellino S, Longo VD. Fasting and cancer: molecular mechanisms and clinical application. Nat Rev Cancer. 2018;18(11):707–19.

Newman JC, Verdin E. Ketone bodies as signalling metabolites. Trends Endocrinol Metab. 2014;25:42–52.

Pietrocola F, et al. Caloric restriction mimetics enhance anticancer immunosurveillance. Cancer Cell. 2016;30:147–60.

Raffaghello L, Lee C, Safdie FM, et al. Starvation-dependent differential stress resistance protects normal but not cancer cells against high-dose chemotherapy. Proc Natl Acad Sci. 2008;105 (24):8215–20.

Sender R, Fuchs S, Milo R. Revised estimates for the number of human and bacteria cells in the body. PLoS Biol. 2016;14(8):e1002533.

Xia S, et al. Prevention of dietary-fat-fueled ketogenesis attenuates BRAF V600E tumor growth. Cell Metab. 2017;25:358–73.

Chapter 16
Do Fasting, CR and FMD Improve the Chemotherapy Response, Reduce Off-Target Toxicities and Enhance Antitumour Immunity? Illusion or Clinical Reality?

Abstract Besides the great strides of efforts made towards understanding the effect of fasting, calorie restriction (CR) and fasting-mimicking diet (FMD), yet there is growing curiosity among the cancer survivors towards this unresolved question whether fasting/CR/FMD has any role to play in the progression and treatment of cancer. Few notable questions being asked by the cancer patients pertain to efficacy of fasting/CR/FMD in arresting the progression of cancer, killing of cancer cells, boosting of the immune system, and whether it has role in improving the effectiveness of currently prescribed chemotherapy and radiation therapy? The sensitivity of cancer cells towards nutrient depletion and their dependency on alternate metabolites has been emerged as a hallmark of cancer. The current preclinical and clinical research finding has demonstrated that fasting, calorie restriction or fasting-mimicking diets induce variety of alterations in growth factors, levels of metabolites, etc., which ultimately leads into creation of an environments that has potential to reduce the adaptive abilities of cancer cells towards survival and thereby helps in improving the efficacy of cancer chemotherapy. Moreover, fasting/calorie restriction or FMDs have been reported to increase resistance towards chemotherapy drugs in normal cells; however, it is not found in cancer cells and thereby promote regeneration effects in normal cells. These dramatic effects of fasting/FMD may help in preventing the detrimental side effects of chemotherapy drugs. Although fasting cannot be tolerated by patients, however, the current preclinical and clinical studies demonstrated feasibility and safety of periodic fasting or FMDs. Taking into considerations the current literature, it is possible to integrate the periodic fasting or FMDs with chemotherapy, cancer immunotherapy or other cancer treatments, which may increase the therapeutic index of cancer treatments and circumvent the emerging resistance and side effects or off-target toxicities.

Keywords Fasting · CR · FMD · Chemotherapy · Off-target toxicity · Antitumour immunity

R. N. Gacche, *Dietary Research and Cancer*,
https://doi.org/10.1007/978-981-16-6050-4_16

Abbreviations

ACC1	Acetyl-coenzyme A carboxylase 1
AFT-3	AMP-dependent transcription factor
AKT	Protein kinase B
CD	Cluster of differentiation
CML	Chronic myeloid leukaemia
COX-2	Cyclooxygenase-2
CR	Calorie restriction
ENT1	Equilibrative nucleoside transporter 1
ET-1	Endothelin 1
FMD	Fasting-mimicking diet
FoxO1	Forkhead transcription factor O class 1
FoxO3	Forkhead transcription factor O class 3
H2AX	H2A histone family member X
HER2	Human epidermal growth factor receptor 2
HIF-1α	Hypoxia-induced factor 1α
HORMAD1	HORMA domain-containing protein 1
ICAM-1	Intercellular adhesion molecule 1
IGF	Insulin-like growth factor
KD	Ketogenic diet
mCRPC	Metastatic castration-resistant prostate cancer.
NF-κB	Nuclear factor-kappa light-chain enhancer of activated B cells
NK	Natural killer cells
NSCLC	Non-small cell lung cancer
PD-1	Programmed cell death protein 1
PI3K	Phosphoinositide 3 kinase
PPAR-γ	Peroxisome proliferator-activated receptor γ
PTEN	Phosphatase and tensin homolog
RAS	Rat sarcoma
SIRT1	Sirtuin 1
TKI	Tyrosine kinase inhibitor
TNBC	Triple-negative breast cancer
ZEB1	Zinc finger E-box-binding homeobox 1

Introduction

The advances in molecular cancer biology has changed our perception towards the pathophysiology, progression and metastasis and also inspired to develop novel, effective, safe and target-oriented anticancer agents. Notwithstanding with cancer alone, the research in cancer biology has enriched our knowledge in general pathophysiology of other human ailments as well. Now, it has been widely accepted that cancer is a multifarious disease, which can initiate by nonmodifiable factors (e.g. genetic instability, ethnicity, gender and age) and by malleable factors such as diet, alcohol consumption, smoking, physical inactivity, obesity, high-fat diet, and exposure to radiations and carcinogenic agents (Brawley 2011; Nindrea et al. 2017; Stein and Colditz 2004).

Fasting, caloric restriction (CR) and food-mimicking diet (FMD) have been projected in the recent past as approaches for lowering the risk of age-linked degenerative diseases such as cancer, cardiovascular diseases and neurodegenerative impairment, and also for extending the life span in general and overall survival of cancer patients in particular. The emerging molecular findings have shown the close association of fasting, CR and FMD with cancer immunosurveillance. Precisely, these nonpharmacological approaches reduce the migration of immunosuppressive regulatory T cells in the tumour microenvironment. Fasting, CR and FMD are described to stimulate the process of autophagy via reduction in acetylation of lysine and thereby enhance sensitivity to immuno- and chemotherapy. Nevertheless, downregulation of insulin-like growth factor 1 (IGF1) and reactive oxygen species (ROS) is also some of the crucial mechanism triggered by fasting, CR and FMD. Moreover, fasting, CR and FMD have also been shown to modulate health by supporting gut microbiota diversity, modifying host metabolism, reducing total cholesterol, lowering diastolic blood pressure, reducing triglyceride levels and elevating morning cortisol level, along with positive effects on body fats, cardiopulmonary fitness, weight, fatigue and weakness, and overall improvement in quality of life. The recent research also pinpoints the significance of fasting, CR and FMD in reducing the metastatic and carcinogenic ability of cancer stem cells, which are generally considered as major culprit for the recurrent tumour formation, relapse and drug resistance (Pistollato et al. 2021).

In series of preclinical model studies, it has been reported that the mutations that activate IGF1R, RAS, AKT or PI3KCA or mutations that inactivate PTEN are described in almost all cancers in humans (Hanahan and Weinberg 2011). These mutational observations in series of cancers inspired to design a hypothesis, which states that fasting, CD or starvation might demonstrate adverse effects in cancer cells (unlike of normal cells) in terms of their potential to sustain against various stress conditions, including chemotherapy. Owing to the presence of aforesaid oncogenic mutations that convergently activate growth-promoting signalling pathways, cancer cells are unable to adjust and adapt efficiently to starvation conditions. Consequently, many types of cancer cells, except normal cells, experience various functional discrepancies, becoming susceptible to toxic agents, including chemotherapeutics. To describe in other words, fasting CR or starvation may lead to a differential stress resistance between cancer and normal cells. The clear message from the above hypothesis is that normal cells tackle the fasting or starvation by downregulating genes assembly associated with cell proliferation, which compels cells to enter a self-maintenance mode that not only sustains them in starvation but also protects them from the damages inflicted by chemotherapy, radiotherapy and other toxic substances.

Role of Fasting, CR and FMD in Cancer Chemotherapy

There are numerous preclinical studies, which advocate the fasting intervention towards increasing the efficacy of chemotherapy drugs (Nencioni et al. 2018). For example, in a preclinical animal model study it was observed that 2 days of fasting effectively extended the survival rate of mice injected with high doses of etoposide in comparison with disease control mice fed with normal diet (Raffaghello et al. 2008). Further studies revealed that downregulation of IGF1 signalling pathway in response to fasting primarily provides protection to primary glia and neurons, but unfortunately provides no protection to glioma and neuroblastoma cells treated with cyclophosphamide and pro-oxidative compounds, and it provides protection to mouse embryonic fibroblasts from doxorubicin. The results of many such studies have clearly demonstrated that IGF1 downregulation is a key underlying mechanism involved in imparting increasing chemotherapy tolerance under fasting conditions (Lee et al. 2010).

While addressing the question, whether fasting improves chemotherapy outcome? The answer to this complicated question is poorly understood; however, the data obtained from preclinical model studies conducted in animal models, including metastatic cancer models, are encouraging. The outcome of such studies clearly showed that periodical fasting or FMD interventions resulted in pleiotropic anticancer effects and improved the therapeutic efficacy of chemotherapy drugs including TKIs along with manifestation of regenerative and protective in multiple organs (Brandhorst et al. 2015). For example, in metastatic neuroblastoma model studies, 48 h of fasting cycles mitigated the high-dose toxicity of etoposide in mice; however, it did not reduce the activity profile of etoposide against neuroblastoma allografts (Raffaghello et al. 2008). In another metastatic cancer animal model studies, 48 h of fasting cycles in combination with doxorubicin or cyclophosphamide was found significantly effective in arresting the subcutaneously growing tumours and also extended the survival of BLLB/c animals in metastatic breast cancer, melanoma and neuroblastoma as compared to solo drug treatment (Lee et al. 2012). In BALB/c mouse colon cancer allograft model studies, 48 h of fasting strengthened the anticancer potential of oxaliplatin, demonstrated anti-Warburg effects and induced apoptosis in cancer cells by promoting oxidative stress (Bianchi et al. 2015). In subcutaneous tumour model (CT26), the 24 h of alternate-day fasting over a period of 2 weeks resulted in inhibition of growth in colon cancer along with reduced production of extracellular adenosine by cancer cells and by downregulating CD73 expression. It has been described that CD73 is involved in conferring the immunosuppressive M2 phenotype to macrophages (Sun et al. 2017). In yet another human pancreatic cancer xenograft model studies, 24 h of fasting before injecting gemcitabine resulted in delayed pancreatic cancer progression along with increased levels of ENT1 in tumours. The described preclinical model studies clearly suggest that periodic fasting approach and intervention of FMD in combination with current chemotherapy have significant potential to improve the therapeutic index of cancer treatment along with reducing the off-target side effects. These studies also inspired to conduct series of human clinical trials (Table 16.2) for unravelling the effect of fasting and FMDs in progression and treatment of variety of cancers.

Preclinical studies have clearly demonstrated that the intervention of CR can successfully shield animals from the deleterious effects of chemotherapy and also enhance the therapeutic potency of various caner chemotherapeutic agents (Lee et al. 2012; Ibrahim et al. 2021). CR has been reported to induce the activation of AMP-activated protein that ultimately leads to heightened apoptosis (Mukherjee et al. 2004). Some cancer patients cannot tolerate the adverse of CR and fasting; for such patients, the agents that mimic CR like metformin, rapamycin and resveratrol may also serve as an effective alternative for the manifestation of protective effects of CR. For example, in preclinical animal model studies, injection of the CR mimetic rapamycin also called as sirolimus showed mTOR inhibition, extended overall survival and also delayed the progression of cancer in mice (Harrison et al. 2009). Similarly, another CR mimetic drug metformin has demonstrated the suppression of tumour growth. Similar effect was also observed in series of experimental models like mammary and colon and also in haematopoietic cancer models (Pollak 2012). There are various health concerns associated with the use of chronic CR in cancer patients, as they are already at a higher risk of weight loss owing to toxic effect of cancer chemotherapy drugs, cachexia and sarcopenia that arise from tumour-derived signals that degrade muscle and adipose. Another physiological concern associated with the anti-inflammatory effect of CR is that it might create adversities in cancer patients who are immunodeficient either because of disease or its therapy (Ibrahim et al. 2021).

Role of Fasting, CR and FMD in Immunosurveillance in Cancer

It is well-accepted fact that unbalanced nutritional conditions, like malnutrition and diet-induced obesity, significantly affect the immunological status of the body. Thus, diet and nutritional status are closely associated with immune response. Fasting and CR without causing malnutrition involve a chronic reduction in energy intake by 15–40% as compared with ad libitum conditions. The scientific link fasting/CR and immunosurveillance were unravelled nearly half a century ago (Fernandes et al. 1976; Most et al. 2017). Since then, a series of preclinical and clinical studies have demonstrated that fasting/CR helps in suppressing the genesis of several human diseases, such as cancer, cardiovascular disease, autoimmune disorders and diabetes (Most et al. 2017; Taylor 2019; Iwashige et al. 2004). It is proved that the activation of effector T (Teff) cells and M1 macrophages is largely contingent on the phosphatidylinositol-3 kinase (PI3K)-Akt-mechanistic target of rapamycin (mTOR) signalling pathway, and consistent activation of this pathway by overnutrition leads to drive M1-skewed inflammation. On the other hand, low-energy status conferred by CR suppresses the PI3K/Akt/mTOR axis along with reciprocal activation of adenosine monophosphate-activated protein kinase (AMPK) and sirtuin family proteins (Mercken et al. 2013; Cantó and Auwerx 2011; Guarente 2013). It is proved that AMPK inhibits the activity of acetyl-coenzyme A carboxylase 1 (ACC1), which ultimately leads to a reduction in fatty acid synthesis (Wakil and Abu-Elheiga 2009). Finally, the altered lipid metabolism determines the T-cell fate decision. For example, a 30% reduction in food intake for a period of 4 weeks will constrain the differentiation into Th17 cells, while promoting the development of regulatory T cells by the inhibition of ACC1 in naive T cells.

Sirtuin 1 (SIRT1) is one of the NAD+-dependent deacetylases and acts as a modulator of immune response and metabolism. Fasting and CR are known to upregulate the expression of SIRT1 protein in multiple cells and tissues (Nemoto et al. 2004; Cohen et al. 2004; Ran et al. 2015). SIRT1 deacetylates NF-κB, PPAR-γ and HIF-1α and regulate the innate and adaptive immune responses. For example, in macrophages, inadequacy of SIRT1 leads to hyperactivation of NF-κB, followed by the overexpression of pro-inflammatory cytokines like TNF-α and IL-1β (Schug et al. 2010). The immunomodulatory effects of fasting and CR are also be manifested via FoxO1 and FoxO3 transcription factors. It has been reported that fasting/CR alleviates PI3K/Akt signalling pathway, which happens to be a negative regulator of FoxO1 and FoxO3. Fasting/CR subsequently upregulates FoxO1 and FoxO3 transcription factors. In CD4+ T cells, FoxO1 and FoxO3 upregulate the expression of Foxp3, the chief regulator of regulatory T (Treg) cells (Kim et al. 2008; Hori et al. 2017; Ouyang et al. 2012; Ohkura and Sakaguchi 2010).

Animal model studies conducted by Di Biase et al. demonstrated the immune beneficial effects of FMD. They demonstrated that when breast cancer cells (4T1) are grafted into BALB/c mice or male C57BL/6 mice grafted with melanoma cells (B16), when provided with only FDM for several cycles or in form of combination along with chemotherapy (doxorubicin and cyclophosphamide) resulted in significant surge of CD3+/CD8+CTLs and a considerable decline in population of CD3+/CD4+/CD25+Treg in the tumour tissue, which ultimately resulted in decrease in cancer progression (Di Biase et al. 2016). The role of OX40 agonist (αOX40) has been reported to mediate cancer immunotherapy by increasing T-cell function and inducing tumour regression in animal model studies (Ruby and Weinberg 2009). In animal model studies, it was observed that CR (but not resveratrol) was able to restore the OX40-mediated antitumour immunity, thereby retaining antigen-specific CD4+ T-cell priming in 12-month-old mice grafted with either sarcoma or breast tumour cells and treated with αOX40 (Farazi et al. 2014).

In other preclinical settings, the effect of fasting and CR has been shown to regulate activity of natural killer (NK) cells, which interestingly recognize both transformed neoplastic cells and virally infected cells. Six-month-old mice, which were undergoing CR, showed reduced levels of mature NK cells in peripheral tissues, and significantly large numbers of CD127+ cells in bone marrow, spleen and lymph nodes, with escalated production of TNF-α and GM-CSF levels, and lowered levels of gamma IFN-γ when stimulated with IL-2, IL-12 or anti-NK1.1 (CD161). Interestingly, the NK cells from CR mice were found more perceptive to YAC-1 T-cell lymphoma cells, with CD27−/CD11b+ NK cells yielding more granzyme B and imparting a higher frequency of degranulation than CD27−/CD11b + NK cells from ad libitum control mice (Clinthorne et al. 2013).

Sizable data have accumulated in the recent years, which suggest that fasting or FMDs alone or to a higher magnitude when combined with chemotherapy activates several lymphoid progenitors and thereby boosts immune attack against tumours via diverse mechanisms (Pietrocola et al. 2016; Sun et al. 2017; Di Biase et al. 2016). The intervention of FMD under in vivo condition downregulated the expression of HO1 in cancer cells, whereas HO1 expression was upregulated in normal cells. The HO1 is a protein, which provides protection against oxidative damage and apoptosis. It was also observed that reduced expression of HO1 in cancer cells imparts FMD-induced chemosensitization by increasing CD8+ tumour-infiltrating lymphocyte-dependent cytotoxicity, which may be associated with the downregulation of regulatory T cells (Di Biase et al. 2016). A recent study conducted in a mouse colon cancer model by alternate-day fasting for a period of 2 weeks demonstrated that fasting activated autophagy in cancer cells and downregulated the expression of CD73, which resulted in lower production of immunosuppressive adenosine by cancer cells (Sun et al. 2017). Ultimately, fasting induced downregulation of CD73 prevented the macrophage transformation into immunosuppressive M2 phenotype. These studies conducted clearly advocate fasting or FMDs can be a complementary intervention in combination with immune checkpoint inhibitors (Postow et al. 2015).

Series of clinical trials of fasting or FMD in cancer patients taking chemotherapy support its tolerability, plausibility and overall safety for cancer patients (Dorff et al. 2016; de Groot et al. 2015). In a small-size (34 patients) randomized clinical trial, it was observed that FMD aided patients in maintaining their quality of life during chemotherapy treatment and also reduced fatigue (Bauersfeld et al. 2018). The preliminary data also suggest that fasting or FMDs aid in minimizing the chemotherapy-triggered DNA damage in normal cells of cancer patients (Dorff et al. 2016; de Groot et al. 2015). The ongoing clinical trials addressing the role of FMDs in cancer patients (ClinicalTrials.gov) will provide more understanding towards combination of periodic FMDs with conventional anticancer drugs and its benefits of cancer survivors. From the research data accumulated in the recent past, it is crucial to realize that the FMDs may not be highly effective in amelioration of side effects of cancer treatments nor it may enrich the efficacy of all cancer therapies; however, it will have great potential to do so possibly for a major fraction of patients and majority of anticancer drugs. It is also important to understand that the weak or malnourished cancer patients or patients who are at high risk of malnutrition must not be enrolled as participants in clinical studies involving fasting or FMDs.

Is There a Valid Clinical Evidence of Fasting or CR Induced Protection Against Cancer?

Evolving epidemiological studies and preclinical animal model studies including humans and monkeys support the notion that periodic fasting, chronic CR and and/or intervention of FMD can be a complementary strategy for integrated management of cancer (Mattison et al. 2012). Based on the preclinical and clinical research outcome, the evidence-based recommendations and guidelines for preferred and nonpreferred foods along with lifestyle suggestions to minimize cancer risk are being acclaimed (Emmons and Colditz 2017; Kerr et al. 2017). The major task is now to identify and standardize a well-tolerated universal periodic fasting, CR or FMD interventions with minimum or no side effects and assess their cancer-preventive potential by conducting human clinical trials. Inspired with the preclinical data, series of human clinical trials have been completed and many are being recruited to assess the effect of fasting, CR or FMD on cancer progression and improving chemotherapy efficacy and tolerance. Until end of June 2021, when a search was carried with a keyword "Fasting and Cancer" at Clinicaltrials.gov, there were over 18 human clinical trials with completed status and approximately 26 studies were under "being recruited" status.

The outcome of series of human clinical trials investigating the effect of periodical fasting/CR/FMD has been published in recent times. For example, in a randomized clinical trial consisting of 100+ patients with HER2-negative stage II/III breast cancer, they received the FMD or their regular diet for 3 days before and during neo adjuvant chemotreatment. No significant difference in toxicity profile in both treated and control groups was observed. Moreover, a per-protocol analysis revealed 90–100% loss of tumour cells in patients who received the FMD. Also, the FMD substantially reduced the chemotherapy-induced DNA damage in T-lymphocytes (de Groot et al. 2020). A phase II FAME trial is designed to investigate the hypothesis that the incorporation of metformin, with or without cyclic FMD, to standard platinum-based chemotherapy enhances the progression-free survival of patients with advanced, LKB-1 inactive lung adenocarcinoma (Vernieri et al. 2019). In a prospective, randomized clinical trial of calorie-restricted KD and fasting in addition to re-irradiation in 20 malignant glioma patients, it was observed that the patients who had forepassed KD-IF developed ketosis by day 6 with significant dropdown in glucose levels. The study concluded that the intervention of KD-IF is feasible and efficacious in inducing ketosis in heavily pretreated patients with recurrent glioma (Voss et al. 2020). Although the clinical trial could not achieve positive outcome of FMD during chemotherapy treatment in breast cancer patients, however, the overall negative outcome of the study highlighted the need for management of future trial design and testing alternative FMD-based therapeutic combinations. The study outcome also provided the baseline information for designing future trials describing that "FMD antitumor activity could be undermined by cancer cells via adaptive metabolic mechanisms allowing them to survive even in glucose and growth factors deprived conditions". It was observed that cancer cells can adapt to fasting/FMD-induced reduction of extracellular glucose by maximum exploitation of mitochondrial oxidative phosphorylation as an alternative source of ATP and anabolic precursors (Vernieri et al. 2019, 2020).

In a clinical trial (NCT03340935), the results revealed that the intervention of three FMD cycles was associated with lowered risk factors and biomarkers of ageing, diabetes, cardiovascular disease and cancer without demonstrating any major detrimental effects (Brandhorst et al. 2015). In an individually randomized crossover trial, testing the effect of short-term fasting on quality of life in patients with gynaecological cancer during chemotherapy, it was observed that the chemotherapy-induced reduction in quality of life was less than the minimally important difference (MID) with short-term fasting but found to be greater than the MID for nonfasted durations. The study revealed no serious adverse effects and concluded that the short-term fasting during chemotherapy can be endorsed and appears to boost the quality of life and minimizes fatigue during chemotherapy (Bauersfeld et al. 2018). A clinical trial was designed to study the safety and feasibility of dose-escalated fasting prior to platinum-based chemotherapy in cancer patients. The outcome of the study reported that the fasting-related toxicities were limited to $\leq$ grade 2, most commonly related to fatigue, headache and dizziness. The DNA damage was significantly reduced in leukocytes of the human subjects who fasted for more than 48 h. IGF-1 levels were observed to reduce by 30, 33 and 8% in the 24, 48 and 72 h of fasting cohorts, respectively, after the first fasting period. The study concluded that the fasting for 72 h around chemotherapy administration can be considered safe and feasible for cancer patients. Biomarkers such as IGF-1 may facilitate assessment of differences in chemotherapy toxicity in subgroups achieving the physiologic fasting state (Dorff et al. 2016).

A clinical trial was designed as pilot study for examining the possibility of short-term fasting (STF) and its effects on chemotherapy tolerance in a patient group (homogeneous) with early breast cancer. The results of the study indicated that the intervention of STF was well tolerated by cancer patients. The mean erythrocyte and thrombocyte counts were substantially higher after 7 days postchemotherapy in the STF group as compared to the non-STF group. The nonhaematological toxicity did not show any major difference between the groups. Moreover, the expression levels of γ-H2AX were substantially upregulated after 30 min postchemotherapy in $CD45^+$ $CD3^-$ cells in non-STF, but not in STF patients. The trial concluded that STF was found to be endorsed well and substantially minimized the haematological toxicity of TAC in HER2-negative breast cancer. Moreover, the study also speculated that the STF might reduce a transient increase in and/or promote a faster recovery of DNA damage in PBMCs after chemotherapy (Dorff et al. 2016). A phase II DIRECT clinical trial of 131 patient of HER2-negative stage II/III breast cancer was designed to study the effect of FMD towards the neoadjuvant chemotherapy as compared to a regular diet. It was observed that the quality of life and the distress scores got reduced during the treatment period and restored to baseline 6 months after surgery. The per-protocol analyses showed better emotional, physical, cognitive and social functioning scores as well as reduced fatigue, nausea and insomnia symptom scores for patients' adherent to the FMD in comparison with nonadherent patients and patients on their regular diet. The study concluded that the intervention of FMD can be an adjunct to neoadjuvant chemotherapy, which improves quality of life and illness perception domains in patients with HER2-negative breast cancer (Lugtenberg et al. 2021). A systematic review of total of 14 studies conducted to assess the effect of fasting on cancer progression observed that homogeneous and robust endpoints are still lacking in the conducted trials. The analysis revealed that there is an urgent need of conducting randomized control trials for evaluation of overall survival and disease-free survival during the treatment of ovarian cancer (Rinninella et al. 2019). The details of human clinical trials (completed and ongoing/ recruited) designed to investigate the effect of fasting/CR/FMD on cancer progression, chemotherapy and toxicity are summarized in Tables 16.1 and 16.2. In summary, the entire discourse on fasting/CR/FMD in concert with cancer clearly suggests that combination of fasting/CR/FMD with chemotherapy, immunotherapy or other treatments modalities may represent a potentially promising integrated cancer management strategy, which perhaps may improve therapeutic index of current chemotherapy drugs, circumvent emerging chemoresistance and might help in amelioration of inscrutable off-target toxicities and known side effects of anticancer drugs.

Conclusion

Dietary interventions have been proven to have a significant impact on body metabolism and disease progression. The sensitivity of cancer cells towards nutrient and energy deficiency or overflow is an evolving area in the mainstream of cancer biology. Emerging preclinical evidence has demonstrated the short-term fasting and CR in amelioration of chemotherapy toxicity, side effects and improvements in efficacy of a variety of cancer chemotherapy drugs in various human cancers. In summary, the entire discourse on fasting/CR/FMD in concert with cancer clearly suggests that combination of fasting/CR/FMD with chemotherapy, immunotherapy or other treatments modalities may represent a potentially promising integrated cancer management strategy, which perhaps may improve therapeutic index of current chemotherapy drugs, circumvent emerging chemoresistance and might help in amelioration of inscrutable off-target toxicities and known side effects of anticancer drugs. The interventions of fasting/CR or FMD should be implemented with at most care by knowing the physiological and immune status of the individual cancer patients, while the global applications of fasting/CR or FMD without knowing the health status, stage and type of treatment employed may lead to creation of complex adversities in cancer patients. Design and conduct of focused, target-oriented randomized clinical trials might be value addition towards the judicious therapeutic utility of fasting/CR/FMD in cancer management.

Table 16.1 Summary of clinical trials completed assessing the effect of fasting/CR/FMD or dietary restrictions in cancer patients

Trial ID	Title of the trial	Trial purpose	Fasting interventions	Responsible party
NCT01954836	"Short-term fasting during chemotherapy in patients with gynecological cancer-a randomized controlled cross-over trial"	"Short-term fasting during chemotherapy in patients with gynecological cancer a randomized controlled cross-over trial"	Short-term fasting, 60–72 h	Andreas Michalsen, Charite University, Berlin, Germany
NCT00936364	"Short-term fasting: impact on toxicity"	Reducing side effects in patients receiving gemcitabine hydrochloride and cisplatin for advanced solid tumours	Short-term fasting, 24–72 h	University of Southern California, Los Angeles, California, USA
NCT01954836	"Short-Term Fasting During Chemotherapy in Patients with Gynecological Cancer-a Randomized Controlled Cross-over Trial (FIT)"	Fasting before (48 h) and 1 day after chemotherapy may protect normal cells from the adverse effects of chemotherapy	Short-term fasting, 60–72 h	Charite University Berlin, Germany
NCT03340935	Safety, Feasibility and Metabolic Effects of the Fasting Mimicking Diet (FMD) in Cancer Patients	Safety of the fasting-mimicking diet (FMD) in cancer patients	NIA	Dean of Medical Oncology and Hematology Department, Fondazione IRCCS Istituto Nazionale dei Tumori, Milano Milan, Italy
NCT01424930	A Study to Determine the Short-Term Safety of Continuous Dosing of Abiraterone Acetate and Prednisone in Modified Fasting and Fed States to Patients With Metastatic Castration-Resistant Prostate Cancer	To establish the safety profile of oral (by mouth) abiraterone acetate and oral prednisone following short-term administration after standardized low-fat or high-fat meals to patients with metastatic (spreading) castration-resistant prostate cancer (mCRPC)	Short-term fasting, 2 h before and 1 h after the dose	Janssen-Ortho Inc., Canada
NCT01304251	Effects of Short-term Fasting on Tolerance to Chemotherapy	To determine the effect of short-term fasting on tolerance to adjuvant chemotherapy in breast cancer patients	Short-term fasting, 24 h before and 24 h after the dose	Leiden University Medical Center Leiden, Netherlands

(continued)

Table 16.1 (continued)

Trial ID	Title of the trial	Trial purpose	Fasting interventions	Responsible party
NCT01802346	"Controlled low calorie diet in reducing side effects and increasing response to chemotherapy in patients with breast or prostate cancer"	Assessments of a low-calorie diet in reducing side effects and increasing response to chemotherapy in patients with breast or prostate cancer	Low-calorie diet 3 days prior to and 1 day after chemotherapy	University of Southern California, Los Angeles, California, USA
NCT01175837	Short-Term Fasting Before Chemotherapy in Treating Patients with Cancer	Fasting before chemotherapy may protect normal cells from the side effects of chemotherapy	Short-term fasting up to 48 h	Mayo Clinic Rochester, Minnesota, United States
NCT02299505	Pharmacokinetic and Safety Study of Lower Doses of Ceritinib Taken with a Low-fat Meal Versus 750 mg of Ceritinib in the Fasted State in Adult Patients With (ALK-positive) Metastatic Non-small Cell Lung Cancer (NSCLC)	Assess the systemic exposure, efficacy and safety of 450 mg ceritinib taken with a low-fat meal and 600 mg ceritinib taken with a low-fat meal as compared with that of 750 mg ceritinib taken in the fasted state in adult patients with ALK rearranged (ALK-positive) metastatic non-small cell lung cancer (NSCLC)	INA	Novartis (Novartis Pharmaceuticals)
NCT00757094	Safety and Feasibility of Fasting While Receiving Chemotherapy	Fasting (refraining from eating and drinking) during the day is safe while receiving chemotherapy	INA	Ali Al-Shanqeeti, King Fahad Medical City Riyadh, Saudi Arabia
NCT01930006	Safety Study of Two Oral Formulations of MGCD265 Administered in Healthy Subjects in the Fasting State	To compare the rate and extent of absorption of two MGCD265 oral formulations at a dose level of 100 mg administered to healthy male and female subjects under fasting conditions	INA	Mirati Therapeutics, Inc. is located in San Diego, CA, United States

NCT02126449	Dietary Restriction as an Adjunct to Neoadjuvant Chemotherapy for HER2 Negative Breast Cancer (DIRECT)	To evaluate the impact of the FMD on tolerance to and efficacy of neoadjuvant chemotherapy in women with stage II or III breast cancer	INA	J.R. Kroep, MD, PhD, Leiden University Medical Center
NCT00959335	Relative Bioavailability Study of Bicalutamide 50 mg Tablet and Casodex Following a 50 mg Dose in Healthy Subjects Under Fasting Conditions	To demonstrate the relative bioavailability of Bicalutamide 50 mg tablet and Casodex following a 50 mg dose in healthy subjects under fasting conditions	INA	Eric Mittleberg, Ph.D, VP of Product Development, Sandoz Inc.
NCT00552955	Effect of Fasting on the Size of Abdominal Lymphatic Tumors in Women	To understand the factors that produce changes in size of lymphangioleiomyomas, as well as to improve the ability of medical professionals to diagnose lymphangioleiomyomas and avoid confusing these tumours with other malignant tumours	INA	National Heart, Lung, and Blood Institute (NHLBI), Maryland, United States
NCT03882281	Pharmacokinetics Profiles of HQP1351 Under Fasting and High-fat Meals in Patients with Chronic Myeloid Leukemia (CML)	To characterize the pharmacokinetics of drug (HQP1351) in participants with resistant chronic myeloid leukaemia (CML) in chronic phase (CP) after high-fat and fasting meals separately	INA	Ascentage Pharma Group Inc., Suzhou, China

INA information not available

Table 16.2 Summary of clinical trials "being recruited" for assessing the effect of fasting/FMD in cancer patient

Trial ID	Title of the trial	Trial purpose	Trial location
NCT04708860	Feasibility of Fasting and Exercise in Pts With HR+ MBC	To explore a combined prolonged nightly fasting (POF) and exercise programme impact on hyperglycaemia and other adverse metabolic consequences of PI3 kinase inhibitors in women with metastatic breast cancer (MBC)	Dana Farber Cancer Institute Boston, Massachusetts, USA
NCT02710721	Fasting and Nutritional Therapy in Patients with Advanced Metastatic Prostate Cancer	To evaluate the effectiveness of intermittent fasting as a supplementary therapy in patients with CRPC or hormone-sensitive prostate cancer with high metastatic load (1≥ visceral and ≥4 osseous metastases) in respect to quality of life, reduction in side effects and possible reduction in tumour progression	Charité Centrum ChirurgischeMedizin, CC 8 KlinikfürUrologie, Berlin, Germany
NCT04387084	Short-term Fasting Prior to PD-1/PD-L1 Inhibitor Therapy for of Advanced or Metastatic Skin Malignancy	To evaluate the safety and feasibility of short-term fasting in combination with PD-1 inhibition therapy for patients with advanced malignancy	University of Southern California, USA
NCT03340935	Safety, Feasibility and Metabolic Effects of the Fasting Mimicking Diet (FMD) in Cancer Patients	To assess the safety, feasibility and metabolic effects of the FMD in cancer patients treated with different standard antitumour therapies	Dean of Medical Oncology and Hematology Department, Fondazione IRCCS Istituto Nazionale dei Tumori, Milano, Italy
NCT04288336	Daily, Long-Term Intermittent Fasting for the Prevention of PSA-Recurrence in Patients With Localized Prostate Cancer After Radical Prostatectomy	To determine whether men treated for localized prostate cancer adhere to a long-term (months–years) daily intermittent fasting regimen	Mayo Clinic, Rochester, Minnesota, United States
NCT03700437	Fasting-mimicking Diet With Chemo-immunotherapy in Non-small Cell Lung Cancer (NSCLC)	To determine the feasibility and compliance of administering a fasting-mimicking diet (FMD) in patients with	Indiana University, USA

(continued)

Table 16.2 (continued)

Trial ID	Title of the trial	Trial purpose	Trial location
		advanced NSCLC receiving maintenance therapy and to assess a patient's willingness to fast for second cycle	
NCT04248998	Calorie Restriction With or Without Metformin in Triple Negative Breast Cancer	The BREAKFAST trial to evaluate the antitumour activity of combining cyclic fasting-mimicking diet (FMD), which reproduces the in vitro effects of glucose starvation, plus/minus metformin with standard preperative anthracycline-taxane chemotherapy in patients with stage I-III TNBC	Fondazione IRCCS Istituto Nazionale Tumori Milano, Italy
NCT03595540	Fasting-mimicking Diet in Patients Undergoing Active Cancer Treatment	To assess the feasibility, safety and effects on patient nutritional status of a 5-day fasting-mimicking diet (FMD) in patients with different cancer types and concomitant anticancer treatment	University of Genova, Italy
NCT00936364	Short-Term Fasting: Impact on Toxicity	To determine the safety and feasibility of short-term fasting prior to administration of combination chemotherapy with platinum in patients with advanced solid tumour malignancies	USC/Norris Comprehensive Cancer Center Los Angeles, California, United States
NCT04027478	Can Fasting Decrease the Side Effects of Chemotherapy?	To assess differences in toxicities in patients undergoing chemotherapy with a combination of taxol/carboplatin when using a fasting-mimicking diet when compared to normal diet before and after treatment	Sutter Cancer Center, Sacramento, California, United States

(continued)

Table 16.2 (continued)

Trial ID	Title of the trial	Trial purpose	Trial location
NCT04461938	Characterization of Metabolic Changes in the Glioma Tumor Tissue Induced by Transient Fasting (ERGO3)	To characterize metabolic and immunological changes in the glioma tumour tissue induced by a fasting cycle of 72 h prior to biopsy or resection	Goethe University/Dr. Senckenbergischen Instituts für Neuroonkologie Frankfurt, Hessen, Germany
NCT03454282	Impact of Dietary Intervention on Tumor Immunity: the DigesT Trial	To assess the immunological and metabolic changes induced by the fasting-mimicking diet (FMD) in the preoperative and postoperative setting in breast cancer and melanoma patients	Fondazione IRCCS Istituto Nazionale dei Tumori, Milan, Italy
NCT04626843	Intermittent Fasting and CLL/SLL	To study how nutrition and dietary behaviours affect cancer patients and to assess how fasting alters the course of cancer or improve immune defences: an alternative way to treat or manage cancer	Eleah Stringer Victoria, British Columbia, Canada
NCT03709147	Metformin Plus/Minus Fasting Mimicking Diet to Target the Metabolic Vulnerabilities of LKB1-inactive Lung Adenocarcinoma	To investigate the efficacy of combining standard-of-care platinum-based chemoimmunotherapy with metformin plus/minus FMD in patients with LKB1-inactive, advanced lung adenocarcinoma	Marina Chiara Garassino Milan, Italy
NCT04292041	Fasting Mimicking Diet in Prostate Cancer and Metabolic Syndrome, a Pilot Study	To evaluate the role of intermittent fasting (fasting-mimicking diet) in prostate cancer patients. The primary end point is metabolic health and the secondary endpoint is quality of life	Medical Director, Department of Radiation Oncology, Galway Clinic

References

Bauersfeld SP, et al. The effects of short-term fasting on quality of life and tolerance to chemotherapy in patients with breast and ovarian cancer: a randomized cross-over pilot study. BMC Cancer. 2018;18(1):476.

Bianchi G, et al. Fasting induces anti-Warburg effect that increases respiration but reduces ATP-synthesis to promote apoptosis in colon cancer models. Oncotarget. 2015;6:11806–19.

Brandhorst S, et al. A periodic diet that mimics fasting promotes multi-system regeneration, enhanced cognitive performance, and health span. Cell Metab. 2015;22:86–99.

Brawley OW. Avoidable cancer deaths globally. CA Cancer J Clin. 2011;61(2):67–8.

Cantó C, Auwerx J. Calorie restriction: is AMPK a key sensor and effector? Physiology. 2011;26:214–24.

Clinthorne JF, Beli E, Duriancik DM, et al. NK cell maturation and function in C57BL/6 mice are altered by caloric restriction. J Immunol. 2013;190(2):712–22.

Cohen HY, Miller C, Bitterman KJ, et al. Calorie restriction promotes mammalian cell survival by inducing the SIRT1 deacetylase. Science. 2004;305:390–2.

de Groot S, et al. The effects of short-term fasting on tolerance to (neo) adjuvant chemotherapy in HER2-negative breast cancer patients: a randomized pilot study. BMC Cancer. 2015;15:652.

de Groot S, et al. Fasting mimicking diet as an adjunct to neoadjuvant chemotherapy for breast cancer in the multicentre randomized phase 2 DIRECT trial. Nat Commun. 2020;11(1):3083.

Di Biase S, Lee C, Brandhorst S, et al. Fasting-mimicking diet reduces HO-1 to promote T cell-mediated tumor cytotoxicity. Cancer Cell. 2016;30(1):136–46.

Dorff TB, et al. Safety and feasibility of fasting in combination with platinum-based chemotherapy. BMC Cancer. 2016;16:360.

Emmons KM, Colditz GA. Realizing the potential of cancer prevention—the role of implementation science. N Engl J Med. 2017;376:986–90.

Farazi M, Nguyen J, Goldufsky J, et al. Caloric restriction maintains OX40 agonist-mediated tumor immunity and CD4 T cell priming during aging. Cancer Immunol Immunother. 2014;63:615–26.

Fernandes G, Yunis EJ, Good RA. Suppression of adenocarcinoma by the immunological consequences of calorie restriction. Nature. 1976;263:504–7.

Guarente L. Calorie restriction and sirtuins revisited. Genes Dev. 2013;27:2072–85.

Hanahan D, Weinberg RA. Hallmarks of cancer: the next generation. Cell. 2011;144:646–74.

Harrison DE, Strong R, Sharp ZD, et al. Rapamycin fed late in life extends lifespan in genetically heterogeneous mice. Nature. 2009;460:392–5.

Hori S, Nomura T, Sakaguchi S. Control of regulatory T cell development by the transcription factor Foxp3. J Immunol. 2017;198(3):981–5.

Ibrahim EM, Al-Foheidi MH, Al-Mansour MM. Energy and caloric restriction, and fasting and cancer: a narrative review. Support Care Cancer. 2021;29:2299–304.

Iwashige K, Kouda K, Kouda M, et al. Calorie restricted diet and urinary pentosidine in patients with rheumatoid arthritis. J Physiol Anthropol Appl Hum Sci. 2004;23:19–24.

Kerr J, Anderson C, Lippman SM. Physical activity, sedentary behaviour, diet, and cancer: an update and emerging new evidence. Lancet Oncol. 2017;18:e457–71.

Kim DH, Kim JY, Yu BP, et al. The activation of NF-κB through Akt-induced FOXO1 phosphorylation during aging and its modulation by calorie restriction. Biogerontology. 2008;9:33–47.

Lee C, et al. Reduced levels of IGF-I mediate differential protection of normal and cancer cells in response to fasting and improve chemotherapeutic index. Cancer Res. 2010;70:1564–72.

Lee C, Raffaghello L, Brandhorst S, et al. Fasting cycles retard growth of tumors and sensitize a range of cancer cell types to chemotherapy. Sci Transl Med. 2012;4:124ra27.

Lugtenberg RT, et al. Quality of life and illness perceptions in patients with breast cancer using a fasting mimicking diet as an adjunct to neoadjuvant chemotherapy in the phase 2 DIRECT (BOOG 2013-14) trial. Breast Cancer Res. 2021;185(3):741–58.

Mattison JA, et al. Impact of caloric restriction on health and survival in rhesus monkeys from the NIA study. Nature. 2012;489:31–321.
Mercken EM, Crosby SD, Lamming DW, et al. Calorie restriction in humans inhibits the PI3K/AKT pathway and induces a younger transcription profile. Aging Cell. 2013;12:645–51.
Most J, Tosti V, Redman LM, Fontana L. Calorie restriction in humans: an update. Ageing Res Rev. 2017;39:36–45.
Mukherjee P, Abate LE, Seyfried TN. Antiangiogenic and proapoptotic effects of dietary restriction on experimental mouse and human brain tumors. Clin Cancer Res. 2004;10:5622–9.
Nemoto S, Fergusson MM, Finkel T. Nutrient availability regulates SIRT1 through a forkhead-dependent pathway. Science. 2004;306:2105–8.
Nencioni A, Caffa I, Cortellino S, Longo VD. Fasting and cancer: molecular mechanisms and clinical application. Nat Rev Cancer. 2018;18(11):707–19.
Nindrea RD, Aryandono T, Lazuardi L. Breast cancer risk from modifiable and non-modifiable risk factors among women in Southeast Asia: a meta-analysis. Asian Pac J Cancer Prev. 2017;18(12):3201–6.
Ohkura N, Sakaguchi S. Foxo1 and Foxo3 help Foxp3. Immunity. 2010;33:835–7.
Ouyang W, Liao W, Luo CT, et al. Novel Foxo1-dependent transcriptional programs control T reg cell function. Nature. 2012;491:554–9.
Pietrocola F, et al. Caloric restriction mimetics enhance anticancer immunosurveillance. Cancer Cell. 2016;30:147–60.
Pistollato F, Forbes-Hernandez TY, Iglesias RC, et al. Effects of caloric restriction on immunosurveillance, microbiota and cancer cell phenotype: possible implications for cancer treatment. Semin Cancer Biol. 2021;73:45–57.
Pollak MN. Investigating metformin for cancer prevention and treatment: the end of the beginning. Cancer Discov. 2012;2:778–90.
Postow MA, Callahan MK, Wolchok JD. Immune checkpoint blockade in cancer therapy. J Clin Oncol. 2015;33:1974–82.
Raffaghello L, et al. Starvation-dependent differential stress resistance protects normal but not cancer cells against high-dose chemotherapy. Proc Natl Acad Sci U S A. 2008;105:8215–20.
Ran M, Li Z, Yang L, et al. Calorie restriction attenuates cerebral ischemic injury via increasing SIRT1 synthesis in the rat. Brain Res. 2015;1610:61–8.
Rinninella E, Fagotti A, Cintoni M, et al. Nutritional interventions to improve clinical outcomes in ovarian cancer: a systematic review of randomized controlled trials. Nutrients. 2019;11(6):1404.
Ruby CE, Weinberg AD. The effect of aging on OX40 agonist-mediated cancer immunotherapy. Cancer Immunol Immunother. 2009;58:1941–7.
Schug TT, Xu Q, Gao H, et al. Myeloid deletion of SIRT1 induces inflammatory signaling in response to environmental stress. Mol Cell Biol. 2010;30:4712–21.
Stein C, Colditz G. Modifiable risk factors for cancer. Br J Cancer. 2004;90(2):299–303.
Sun P, et al. Fasting inhibits colorectal cancer growth by reducing M2 polarization of tumor-associated macrophages. Oncotarget. 2017;8:74649–60.
Taylor R. Calorie restriction for long-term remission of type 2 diabetes. Clin Med (Lond). 2019;19:37–42.
Vernieri C, Signorelli D, Galli G, et al. Exploiting fasting-mimicking diet and metformin to improve the efficacy of platinum-pemetrexed chemotherapy in advanced LKB1-inactivated lung adenocarcinoma: the FAME trial. Clin Lung Cancer. 2019;20(3):e413–7.
Vernieri C, Ligorio F, Zattarin E, et al. Fasting-mimicking diet plus chemotherapy in breast cancer treatment. Nat Commun. 2020;11(1):4274.
Voss M, et al. ERGO2: a prospective, randomized trial of calorie-restricted ketogenic diet and fasting in addition to reirradiation for malignant glioma. Int J Radiat Oncol Biol Phys. 2020;108(4):987–95.
Wakil SJ, Abu-Elheiga LA. Fatty acid metabolism: target for metabolic syndrome. J Lipid Res. 2009;50:S138–43.

Chapter 17
Future Prospectus of Dietary Interventions in the Integrated Management of Cancer

Abstract Series of diet- and nutrition-related questions raised by cancer survivors are waiting to be addressed scientifically. Diet and nutritional recommendations and guidelines suggested by worldwide agencies are either inconsistent or lack the mechanistic and rigorous clinical evidence. Patient-oriented and cancer-specific dietary recommendations are limited. Moreover, the impact of dietary recommendations is not effective owing to moderate interest of clinical oncologists in diet and nutritional interventions. The evolving preclinical and clinical research suggests that the survival of cancer patients is closely associated with dietary interventions, pattern of diet, fasting, calorie restriction and role of dietary ingredients in the pathophysiology of cancer progression. The current landscape of dietary research shows that the diet and dietary ingredients interact with cancer cell proliferating signalling pathways, epigenetic factors, metastasis, cancer stem cells, tumour suppressor or oncogenic miRNAs, and anticancer drug targets. Although a sizable volume of research has been carried out in the diet–cancer area, however, still diet-mediated molecular mechanisms proving benefits to cancer patients are poorly understood and there lies an opportunity of unravelling the diet-mediated inscrutable molecular mechanisms. Some nonscientific illusions about the risk and benefits of dietary supplements (especially antioxidants and vitamins) prevailing in the minds of cancer patients still waiting to be addressed in the mechanistic framework along with launching evidence-based diet-related awareness programmes. Looking forward, the most upcoming discourse may be to investigate the dynamic interplay between dietary interventions and cancer microbiome. There is ample research opportunity to understand the cancer-specific role of human microbiome. Moreover, the future research opportunities in understanding the role of dietary intervention in improving cancer immunosurveillance may be useful in integrating these nonpharmacological regimens with cancer immune therapy.

Keywords Diet · Nutrition · Dietary guidelines · Cancer · Epigenetics · Metastasis · miRNA · Cancer stem cells · Functional foods · Nutraceuticals · Antioxidants · Food supplements · Chemotherapy · Fasting · Overall survival

R. N. Gacche, *Dietary Research and Cancer*,
https://doi.org/10.1007/978-981-16-6050-4_17

Introduction

"What should I eat before and after cancer treatment?" is the most common question asked by cancer patients for which we do not have a strong evidence-based convincing answer. In fact, the answer is complex and is dependent upon several factors such as cancer genetics, treatment prescribed by oncologists, immune status of patients, stage of cancer and age of cancer patient. The advances in preclinical and clinical research have crossed the general notion of "eat healthy-be fit-reduce cancer risk" and a new paradigm of evidence-based personalized dietary interventions are evolving. Overall, the diet and nutrition management is coming up as a complementary or adjuvant therapeutic approach for the effective amelioration of cancer. The cancer survivors have altered physiology owing to the serious side effects and off-target toxicities of chemo- and radiotherapy regimes. Moreover, the toxic effects of therapeutic regimes are more devastating and worst in malnourished or undernourished cancer survivors; therefore, the convincing dietary and nutrition guidelines before, during and after treatment need to be formulated in concert with different types of cancer and pathophysiology of cancer progression. There cannot be a global answer to all cancer survivors. In the recent years, the treatment modalities are shifting from general to personalized. In fact, with the development of personalized medicine approach, there is need to coevolve the personalized diet for cancer patient. The growth and proliferation of tumours depend on supply of nutrients from host, which is basic requirement for the growth and survival of human. Alterations to the host's diet pattern can regulate nutrient availability in the tumour microenvironment. Engineering the tumour microenvironment is one of the evolving research areas for the effective management of cancer growth, and this strategy also represents a promising strategy for arresting the tumour growth. The strategic planning of designing anticancer diet may limit tumour-essential nutritional requirements, and certain dietary elements may target the metabolic activities of the tumour and may also enhance the therapeutic efficacy of the chemotherapy anticancer drugs. The current circumstantial literature that advocates the judicious modification of variety of nutrients in the diet can improve the efficacy of cancer therapeutic approaches. In the current state of the art, while maintaining the healthy status of cancer patients through healthy diet, it is important to consider whether the dietary interventions have any role to play in the framework of tumour/cancer metabolism and thereby affecting the cancer landscape by modulating variety of molecular underpinnings like cell proliferating signalling pathways, epigenetic factors, metastasis, cancer stem cells, tumour suppressor or oncogenic miRNAs and interaction with variety of anticancer drug targets. The entire diet–cancer discourse warrants to consider the dietary habits and nutritional status of a patient in concert with cancer therapy and management.

Need for Developing Conclusive Evidence-Based Dietary Guidelines

Besides sizable progress has been in understanding the impact of diet on cancer risk and prevention, still there are many questions waiting to be answered with concrete, consistent and conclusive clinical evidence of dietary interventions in cancer management. The present dietary recommendations provided by different national and international agencies basically are of "one-fits-all" nature. Perhaps, there is need to shift focus towards personalized cancer-specific diet/nutrition or dietary patterns, which will be in tune with personalized medicine. Recent cancer-specific dietary recommendations of WCRF-AICR described in the CUP report of 2018 (https://www.wcrf.org/dietandcancer/recommendations-about) is a step towards creating a platform for personalized dietary interventions. The evolving research also supports the feasibility and effectiveness of personalized dietary approaches over conventional generic strategies. For example, the outcome of Food4Me RCT conducted for European adults has clearly outlined that personalized nutritional advice delivered by Internet sources had a louder and more appropriate impact on changing the dietary habits as compared to advice delivered by conventional approaches (Celis-Morales et al. 2017). Moreover, the dietary experts also give emphasis on developing patient-centred nutritional recommendations and also states that "Currently, we are not in a position to make personalized recommendations based on sound scientific evidence" (Zick et al. 2018). While formulating the dietary recommendations for cancer survivors, the experts of third WCRF/AICR, CUP Expert Report 2018, have identified six areas of research that needs to be prioritized in the upcoming research endeavours in the mainstream of dietary research. The expert states that there is need to conduct research in areas like (1) "Biological mechanisms by which diet, nutrition and physical activity affect cancer processes"; (2) "The impact of diet, nutrition and physical activity throughout the life course on cancer risk"; (3) "Better characterization of diet, nutrition, body composition and physical activity exposures"; (4) "Better characterization of cancer-related outcomes"; (5) "Stronger evidence for the impact of diet, nutrition and physical activity on outcomes in cancer survivors"; and (6) "Globally representative research on specific exposures and cancer" (WCRF/AICR, Continues Update Project Expert Report 2018; available online at dietandcancerreport.org). The upcoming clinical trials in diet–cancer should customize the endpoint parameters by revisiting the research areas identified by WCRF/AICR, CUP experts. The future drafting of dietary guidelines by different agencies is expected to give more emphasis on evidence-based recommendations. Moreover, there is also a need of national policy or concrete programmes for integrating the dietary recommendations in clinical practice where oncologists have a major role to play. There is need to extend this area of research as the present guidelines are consistently inconsistent in nature.

The New Research Windows in Diet–Cancer Discourse

The sequencing of the human genome has significantly increased our understanding of the human genetic variations and inspired the biomedical research in several aspects. With the advancement in dietary research, the human genome data can be explored for developing personalized nutrition. Moreover, the advances in genetic studies have suggested series of approaches that may contribute towards development of precision diet/nutrition and may provide solutions to emerging problems in public health and may reduce the risk and prevalence of nutrition-related ailments including cancer. The current scientific era is full of "omics" approaches that are employed for unravelling the secrets of molecular mechanisms in various disciplines of biology. It is well-established fact that the individuals differ in their capabilities of metabolizing food and nutrients; therefore, the applications of omics approaches such as nutrigenomics, epigenomics, transcriptomics, proteomics, food omics, metagenomics and metabolomics might strengthen the scientific claims that are required for developing personalized dietary modalities and also for discovery and validation of robust biomarkers of diet in cancer prevention. Perhaps such endeavours of omics-directed (mostly nutrigenetics and nutrigenomics) formulation of personalized dietary/nutritional guidelines may establish a solid foundation for dietary recommendations to different types of cancer survivors. The outcome of such studies may also create grounds for developing personalized novel "epigenetic diet" and many other novel dietary patterns (Ferguson et al. 2016; Lewis and Tollefsbol 2017; Nasir et al. 2020).

The prospective evolving research window in the mainstream of diet–cancer discourse is the linking of dietary interventions with human microbiome owing to its impact on host immunity and response to treatment modalities. For example, the recent report describes the role of gut microbiome in modulating the response to anti-PD-1 immunotherapy in cancer (Gopalakrishnan et al. 2018). The past decades of exponential research driven by high-throughput sequencing technologies coupled with metagenomics (16S sequencing) has widened our understanding of the role of microbiome in human health and pathophysiology of various human ailments including cancer (Valdes et al. 2018). The microbiome–cancer research is being conducted with an inspiration of modulating the microbiome for improving the therapeutic response in cancer (McQuade et al. 2019). The link of dietary inclusions with microbiome is based on a simple hypothesis that the bacteria that harbour in our gastrointestinal tract eat same what we eat. They change us and we change them in a dynamic fashion. There is mounting evidence, which states that the gut microbiome modulates the response to cancer treatment (Geller et al. 2017) and such reports has inspired nutrition research in cancer patients, with a hope of developing gut microbiome as a modifiable target through dietary interventions, which perhaps will improve cancer therapy and prognosis. The gut microbiome's evolving role in personalized cancer medicine is shaping new priorities and avenues for dietary research in cancer patients. For example, dietary inclusion of navy bean or rice bran was found to be associated with alterations in the metabolism of gut microbiome in colorectal cancer survivors (Sheflin et al. 2017). The animal model study demonstrated that dietary fibre has protective role against colorectal tumorigenesis and is dependent on gut microbiome and butyrate (Donohoe et al. 2014). Interestingly, the bacteria *Fusobacterium nucleatum* has been found to be involved in promoting chemoresistance to colorectal cancer by regulating autophagy (Yu et al. 2017). The upcoming research indicates that there is a ample scope for establishing the dietary modulated effects in relation to role of microbiota in carcinogenesis, the impact of gut microbiome on immunity and response to cancer treatments, establishing microbiome as a biomarker in cancer therapy, etc. (McQuade et al. 2019; Daniel and McQuade 2019).

Till date, FDA has not approved any dietary intervention for cancer prevention. Oncologists, clinicians, dieticians and professional cancer caretaker teams are expected to encourage evidence-based dietary recommendations for cancer survivors. Developing legal regulatory framework for the functional foods, nutraceuticals and dietary supplements (DSs) is one of the challenges at national levels, and harmonization of the regulatory affairs at international level is a global challenge as majority of the regulations primarily reflect domestic needs and pressures. Any policy framework for regulation of any consumer product should maintain a balance between consumer benefit, access, informed choice and safety. In the current situation, there is no global consensus on categorizing and defining the products which per say known as DSs, natural health products, food supplements or complementary medicines (Dwyer et al. 2018). The market entry requirements and overall regulatory affairs for these supplemented products vary from country to country. For example, in the USA, neither FDA approval nor registration of the nutraceuticals is required for the sale and the manufacturer is held responsible for its safety, while FDA is empowered for quality surveillance activity. Conversely, in Canada regulations are more stringent, as nutraceuticals are treated quite similar to that of drugs (Shao 2017). Perhaps the regulatory framework for DSs needs to be considered on priority owing to its fast-growing market. There is a global need of framing regulatory framework for protecting the consumers from adverse effects of DSs by having ability to make informed choices about the DS usage. The countries with overlapping cultures, legal systems and economic growth have different regulations in relation to DSs. In countries like the USA, Australia and Canada, DSs are considered appropriate for self-prescription without the intervention or prescriptions from medical practitioner. The key regulations are required to support the evidence-based support claims, legal framework ensuring safety, quality and cost recovery provisions (Binns et al. 2018). Designing educational and awareness programmes for judicious usage of DSs and relying more on dietary sources are essential but practically seem difficult owing to the presence of huge network of industry sector earning massive profits. Perhaps, stringent legal regulatory framework might act as a watchdog over "Profit Rich-Benefit Poor" market of DSs.

The current preclinical and clinical research finding has demonstrated that fasting, calorie restriction (CR) or fasting-mimicking diets (FMDs) improve the efficacy of cancer chemotherapy and ameliorate the side effects and toxicities of chemotherapy drugs. Also, it is well-accepted fact that unbalanced nutritional conditions, like malnutrition, diet-induced obesity significantly affect the immunological status of the body. Thus, diet and nutritional status are closely associated with immune response. In summary, the entire discourse on fasting/CR/FMD in concert with cancer clearly suggests that combination of fasting/CR/FMD with chemotherapy, immunotherapy or other treatments modalities may represent a potentially promising integrated cancer management strategy, which perhaps may improve therapeutic index of current chemotherapy drugs, circumvent emerging chemoresistance and might help in amelioration of inscrutable off-target toxicities and known side effects of anticancer drugs.

Emerging Research Opportunities in Mainstream of Diet–Cancer Area

Looking forward towards the multidimensional diet-mediated effects, the most upcoming discourse will be to investigate the dietary intervention in regulating the dynamics of cancer microbiome. In the midst of evolving human microbiome projects across the world, it has been described that the human body has more micro-organism than human cells (Sender et al. 2016).

The past decades of research inspired the research community to infer the disease biology with human microbiome owing to groundbreaking advances in the field of microbiome research. Sizable literature is available, which describes the inscrutable links of the dynamics of microbiome with pathophysiological complexities of tumour biology. It has been demonstrated that the microbiome plays a role in tumour initiation and progression and influences directly on the tumour cells and indirectly through the manipulation and regulation of the immune system. The recent literature also showed that dynamics of microbiome also determines the response to cancer therapies and predicts disease progression and patient survival. Moreover, modulation of the microbiome can be utilized to explore the efficacy of immunotherapies and for amelioration of drug toxicity (Jain et al. 2021). The logic of linking dietary interventions with microbiome lies in the requirement of host-derived nutrients to the microbial community and understanding their effects on cancer progression and treatments. It is interesting to note that the microbiome is inextricably integrated into the cancer–immune crosstalk and coordination. The microbes are known to interact with the host and the tumour microenvironment in a diverse and context-specific manner, conferring individual and tumour specificities to the immunomodulatory phenotype observed. Unravelling the unique molecular mechanisms through which these alterations are manifested may help in designing targeted and personalized approaches for cancer diagnosis, treatment and prevention (Fessler et al. 2019). Interestingly, the era of immuno-onco-microbiotics is extremely exciting and the excitement can be increased many folds by studying the diet/metabolite-induced regulation of cancer-associated microbiome. Moreover, the research opportunities in understanding the role of dietary interventions in improving cancer immunosurveillance and reducing the migration of immunosuppressive cells (especially regulatory T cells) towards the tumour microenvironment may be useful in integrating these nonpharmacological regimens with cancer immune therapy.

Conclusion

A cursory look at the different dimensions of diet–cancer discourse clearly outlines certain key issues about the inconsistency in dietary recommendations for cancer survivors. The current recommendations are mostly based on limited nonconclusive evidence, and there is lack of strong convincing evidence. There are pros and cons associated with different dietary patterns while employing them as cancer-preventing strategy. Series of national and international agencies and organizations are helping the cancer survivors in providing valuable dietary and other information related to management of cancer. There is evolving trend of evaluating the dietary interventions by conducting RCTs, cohort, prospective and observational studies; however, there is no concrete and conclusive clinical evidence of cancer prevention by dietary interventions. Plethora of preclinical and clinical literature is accumulating in logarithmic fashion unravelling the molecular mechanisms of dietary interactions with variety of pathophysiological processes of cancer progression and anticancer drugs as well. Besides the poor clinical outcome, the market of functional foods, nutraceuticals and dietary supplements is growing exponentially. The current research clearly warns the judicious usage of dietary supplements in cancer amelioration. More precisely, the dietary antioxidant supplements have been linked with increase in certain types of cancer incidence. There is need to mitigate the "Profit Rich-Benefit Poor" market of dietary supplements and launch awareness programmes for educating the cancer survivors. The development of mechanistic and evidence-based dietary recommendations, linking the dietary interventions with microbiome and need of developing a legal framework for safety, quality, efficacy and marketing of functional foods, nutraceuticals and dietary supplements in cancer prevention, is some of the future prospectuses of diet–cancer discourse. More mechanistic framework of preclinical and clinical research is needed for integrating the fasting/CR/FMD in cancer chemotherapy and immunotherapy clinical practices. Looking forward, the most upcoming discourse may be to investigate the dynamic interplay between dietary interventions and cancer microbiome. There is ample research opportunity to understand the cancer-specific role of human microbiome. Moreover, the future research opportunities in understanding the role of dietary intervention in improving cancer immunosurveillance may be useful in integrating these nonpharmacological regimens with cancer immune therapy.

As a part of community awareness, community organizations and public and private agencies across the world should join hands together for the establishment of collaborative network at international, national, state and even at local levels for the development of state-of-the-art cancer patient health and treatment-related SOPs and should develop a programme for its implementation. The collaborative networks should also develop policies and suggest environmental changes that may increase the access to affordable anticancer treatment regimes, nutritious foods, should strive for safe, use friendly, enjoyable and easy-to-access opportunities for physical activities, create active living environments and develop programmes for discouraging the access of alcoholic beverages for all individuals including cancer patients. "ASC guidelines for diet and physical activity for cancer prevention" has also given a strong emphasis on developing innovative national policies by the state agencies for the welfare of cancer patients and normal individuals (Rock et al. 2020). As a part of making national policies by the state agencies, it is time to revisit the policies and frame newer policies, which will provide easy access to affordable healthy foods, develop community food retail strategies, make radical improvements in policies for healthy eating in concert creating active living environments and frame policies for the type of offered by restaurants, re-strengthening the policies towards the importance of physical education and activity in schools and colleges, making policies and building environments for increasing easy access to physical activity facilities for playing grounds, places for leisure time activity, etc., making policies towards decreasing the access to alcoholic beverages (irrespective of tax revenue generation) and developing strategic clinical policies for promoting the healthy eating and active living along with limiting the intake of alcohol.

References

Binns CW, Lee MK, Lee AH. Problems and prospects: public health regulation of dietary supplements. Annu Rev Public Health. 2018;39:403–20.

Celis-Morales C, Livingstone KM, Marsaux CF, Macready AL, Fallaize R, O'Donovan CB, et al. Effect of personalized nutrition on health-related behaviour change: evidence from the Food4Me European randomized controlled trial. Int J Epidemiol. 2017;46(2):578–88.

Daniel CR, McQuade JL. Nutrition and cancer in the microbiome era. Trends Cancer. 2019;5 (9):521–4.

Donohoe DR, Holley D, Collins LB, et al. A gnotobiotic mouse model demonstrates that dietary fiber protects against colorectal tumorigenesis in a microbiota- and butyrate-dependent manner. Cancer Discov. 2014;4:1387–97.

Dwyer JT, Coates PM, Smith MJ. Dietary supplements: regulatory challenges and research resources. Nutrients. 2018;10(1). pii: E41.

Ferguson LR, De Caterina R, Görman U, Allayee H, Kohlmeier M, Prasad C, et al. Guide and position of the International Society of Nutrigenetics/Nutrigenomics on personalised nutrition: part 1—fields of precision nutrition. J Nutrigenet Nutrigenomics. 2016;9(1):12–27.

Fessler J, Matson V, Gajewski TF. Review exploring the emerging role of the microbiome in cancer immunotherapy. J Immunother Cancer. 2019;7(1):108.

Geller LT, Barzily-Rokni M, Danino T, et al. Potential role of intratumor bacteria in mediating tumor resistance to the chemotherapeutic drug gemcitabine. Science. 2017;357:1156–60.

Gopalakrishnan V, Spencer CN, Nezi L, et al. Gut microbiome modulates response to anti-PD-1 immunotherapy in melanoma patients. Science. 2018;359:97–103.

Jain T, Sharma P, Are AC, et al. New insights into the cancer-microbiome-immune axis: decrypting a decade of discoveries. Front Immunol. 2021;12:622064.

Lewis KA, Tollefsbol TO. The influence of an epigenetics diet on the cancer epigenome. Epigenomics. 2017;9(9):1–3.

McQuade JL, et al. Modulating the microbiome to improve therapeutic response in cancer. Lancet Oncol. 2019;20:e77–91.
Nasir A, Bullo MMH, Ahmed Z, Imtiaz A, Yaqoob E, Jadoon M, et al. Nutrigenomics: epigenetics and cancer prevention: a comprehensive review. Crit Rev Food Sci Nutr. 2020;60(8):1375–87.
Rock CL, Thomson C, Gansler T, et al. American Cancer Society guideline for diet and physical activity for cancer prevention. CA Cancer J Clin. 2020;70(4):245–71.
Sender R, Fuchs S, Milo R. Revised estimates for the number of human and bacteria cells in the body. PLoS Biol. 2016;14(8):e1002533.
Shao A. Global market entry regulations for nutraceuticals, functional foods, dietary/food/health supplements. In: Developing new functional food and nutraceutical products. 2017. p. 279–90.
Sheflin AM, Borresen EC, Kirkwood JS, et al. Dietary supplementation with rice bran or navy bean alters gut bacterial metabolism in colorectal cancer survivors. Mol Nutr Food Res. 2017;61.
Valdes AM, Walter J, Segal E, Spector TD. Role of the gut microbiota in nutrition and health. BMJ. 2018;361:k2179.
Yu T, Guo F, Yu Y, et al. Fusobacterium nucleatum promotes chemoresistance to colorectal cancer by modulating autophagy. Cell. 2017;170:548–63.e16.
Zick SM, Snyder D, Abrams DI. Pros and cons of dietary strategies popular among cancer patients. Oncology (Williston Park). 2018;32(11):542–7.